The VHF "How To" Book
A Guide For All Amateurs

The VHF "How To" Book
A Guide For All Amateurs

By Joe Lynch, N6CL

CQ Communications, Inc.

Library of Congress Catalog Card Number 94-71290
ISBN 0-943016-07-X

Editor: Terry Littlefield, KA1STC
Managing Editor: Gail M. Schieber
Layout and Design: Edmond Pesonen
Cover photo: Larry Mulvehill, WB2ZPI
On the cover: Allen Katz, K2UYH
                Robbinsville, New Jersey

Published by CQ Communications, Inc.
76 North Broadway
Hicksville, New York 11801 USA

Printed in the United States of America.

Acknowledgements

I would like to thank my editor, Terry Littlefield, KA1STC, the editor of *Communications Quarterly,* for taking on the task of shaping up this book. I would also like to thank my best friend, Carol King, K5CPZ, who as a high school teacher helps me find most of my split infinitives. My sincere appreciation to VHF+ experts Emil Pocock, W3EP, and Michael Owen, W9IP, for agreeing to peer review the book. As my friend, Bob Cerasuolo, WA6IJZ, the editor of the "West Coast VHFer" says, "They are two of the best!" I'd also like to thank Tom O'Hara, W6ORG, for reviewing the chapter on ATV.

I owe a sincere debt of gratitude to Alan Dorhoffer, K2EEK, the editor of *CQ* magazine, for taking a chance on an unknown writer a few years back when he hired me to be the VHF editor for the magazine. Also, many thanks to Gail Schieber, the managing editor of *CQ* magazine, for all her hard work producing my monthly column. In addition, I'd like to thank my friend Ted Goldthorpe, WA4VCC, who has given me many column ideas over the years.

Another thank you goes to the American Radio Relay League in Newington, Connecticut. Information on the ARRL QSL bureaus, the ARRL awards rules, and the grid locator map appears in this book with their permission.

But most of all, I'd like to thank you, the VHF+ operator, who takes the time to send me a report of your activity or provide a correction for the occasional error in my column, or calls me on the phone or on the air with some words of encouragement. You make the incredible difference between ordinary ham radio and extraordinary communications. Without you, there would be no wonderful world of VHF+ to write about. Thank you ever so much.

Joe Lynch, N6CL

Introduction

Although many of you may already know me from reading my "VHF Plus" column in *CQ* magazine, I'd like to introduce myself to those of you who don't. I've been *CQ*'s VHF editor since 1991. I hold an Extra class license, have been a ham for over 33 years, and was first licensed as WV6PDE in Bonita, California. I also hold an FCC general radio telephone license (the old first phone). You'll find me on all amateur frequencies from 160 to 2 meters, plus 70 cm. For the past five plus years, I've been the ARRL's Oklahoma Section Manager. I'm also the editor of the *QCWA Journal*, a magazine published by the Quarter Century Wireless Association—an organization of hams licensed at least 25 years ago. Because I hold these positions, I attend many hamfests and club meetings, where I keep my finger on the pulse of the amateur radio fraternity.

Throughout this book I use the term "VHF+." There are two reasons for doing so. First, it is the abbreviated way to refer to the VHF spectrum (30–300 MHz) and all the frequencies above, such as ultra high frequency (UHF, 300–3,000 MHz), super high frequency (SHF, 3–30 GHz), extremely high frequency (EHF, 30–300 GHz), and light (above 300 GHz). Second, it is an abbreviated reference to my column in *CQ* magazine, and thus a subtle way for you to think of it when you are looking for the best coverage of VHF and above activities.

My exposure to the VHF+ frequencies goes back to my early days in ham radio, when I spent many nights hanging around the shack of Bert Adams, K6BTO, in Bonita, California. Bert had every imaginable piece of equipment from every radio surplus store in San Diego County. I remember how much it meant to him to be able to modify a piece of gear so he could talk clear across town. I remember the smell of that surplus equipment as the components baked from the heat of those old tubes. I remember the constant hiss of the 1296 MHz receiver (an old APX-6) with its gain kept wide open, so as not to miss a call (and the high pitched buzz of the Navy shipboard radars as they swept across the front end of the receiver). I also remember Bert's dreams of being able to talk just a little bit farther on those incredibly high frequencies.

I asked him why he bothered operating on the high frequencies when he could get on the low bands any time he wanted and talk anywhere in the world. Bert replied that wasn't always the case. He said that in ham radio's early days, radio operators couldn't talk very far because they didn't know how. Eventually they found a way to increase their range, and now it was up to him and other radio pioneers to discover how to use these higher frequencies effectively. I told him it seemed an incredible amount of work just to find out how to communicate across town, and that maybe there just wasn't any way to work anyone much farther away. Bert brushed off my negative comments.

Instead, Bert regaled me with tales of bouncing signals off the Moon, or talking to other stations on 2 meter meteor scatter, or seeing a fellow ham on amateur television. As a newly licensed amateur, I could never quite understand the intrigue such etherial forms of ham radio communication held for him.

I found that the 6 meter ham band held a fascination for some amateur radio operators. When a ham got a Technician class license, he would invariably get on 6 meters. In those days, the Technician didn't have Novice privileges unless he held a Novice license simultaneously (similar to today's straight Technician class licensee who hasn't taken the code test), so 6 meters was the only band that provided long distance communications with any regularity for this license class.

I also noticed something rather peculiar about the Technicians active on 6 meters. They actually enjoyed it there! Then I noticed something else. Some Technicians who upgraded to the General class license stayed on 6 meters. It seemed there was some mystical charm to communications on that band.

Finally, I observed the excitement of 2 meter operators who found they could hear Los Angeles and sometimes Santa Barbara stations very clearly in San Diego on certain nights at certain times of the year. I would marvel to myself about my feelings of indifference toward such seemingly trivial forms of communications.

However, over the years I've discovered that there *is,* indeed, something almost mystical about communications on those higher frequencies. There's something unique about being able to capture snatches of communications with distant stations on frequencies that normally aren't conducive to long distance propagation. I guess, for me, it's a way of cheating nature momentarily. It's a challenge to lure that bit of a signal down my antenna and coax and into my receiver long enough to obtain the vital information needed for a QSO, and to reverse the process by sending my signal back into that finicky atmosphere with the hopes that the operator on the other end will be able to receive it and transcribe the information necessary for a complete contact.

Why did I write this book? When I began venturing into the world of VHF+, I had to consult several volumes to find all I needed to know. I read one book for information on propagation, another book to find out about equipment design, and yet another book to learn frequency allocations and band plans. I had to consult more books to learn about the fun of hidden transmitter hunting and sending signals through satellites. However, none of the books I found told me the protocol of how to communicate over the local repeater.

In this volume, I've combined basic information on the VHF+ frequencies for use as an easy reference for beginners, and as a resource for those old timers who want to learn about operating on these frequencies.

This book was written for you, to provide you with the tools necessary to operate the VHF+ frequencies. I hope you enjoy operating on these bands as much as I have enjoyed writing about them.

Joe Lynch, N6CL

*"The loss of wonder, of awe, of the sense of the sublime,
is a condition leading to the death of the soul."*
—Edmund Fuller

Table of Contents

Part I
VHF+ Operations

Chapter 1

Welcome To The VHF+ Frequencies

Welcome to the VHF+ frequencies. Perhaps you've chosen these frequencies because the amateur radio license you hold doesn't permit you to operate on the HF ham bands. Maybe you're tired of conditions on HF, and are looking for a refuge. Possibly, you just want to try something new. Whatever the reason, you'll find operating challenges that will more than satisfy you.

The hams you meet on the VHF+ frequencies are among the friendliest in the hobby—and the most knowledgeable. Hams who operate on these bands operate on the cutting edge of communications technology and design. The vast majority of them share their knowledge freely. In fact, some say that the VHF+ operator is a throwback to what ham radio used to be—friendly, courteous, and helpful.

The VHF+ frequencies are all those above 50 MHz. If you hold a Technician class license or greater, you are permitted to operate on any of them, using any mode of operation authorized for that particular frequency. Even if you are a Novice class licensee, you can still operate on two of the VHF+ ham bands.

There are many options available for the ham who chooses to work the VHF+ frequencies. Some amateurs work DX on 6 meters. Others operate Earth-Moon-Earth (EME), bouncing signals off the Moon on 6 or 2 meters or 135, 70, 33, 23, 13, 9, 5, or even 3 cm. Still others send signals to closer, manmade satellites. Those with a flair for the dramatic might like to try amateur television, or ATV. Less adventurous operators may simply choose to talk to area hams through the local repeaters from their home QTH, or while mobile.

VHF+ operation presents many challenges to contesters. If contesting is your passion, there are several contests throughout the year. For you award hunters, there are awards specifically tailored for these bands.

If you've always dreamed of operating a "rare" DX station like those found on the low bands, you can become one by operating Rover and traveling to rare grid locators. And, as in other areas of amateur radio, there's always the challenge of building your own equipment for the various bands you plan to operate.

This book is divided into two parts. Part I covers overall operation on the VHF+ frequencies; Part II covers weak signal operating techniques.

A couple of notes on abbreviations: The initials "ARRL" or the word "League" refers to the American Radio Relay League, the national organization for ham radio operators in the United States. The initials "CSVHF" refer to the Central States VHF Society, one of the largest VHF+ organizations in the country.

At various places within this book, I urge you to contact your local guru for more information about a particular type of operation. A mentor can be a real help if you have questions. I was lucky in that mine, Bert Adam, K6BTO, was father to my friend Frank, AE6L.

How do you find a mentor of your own? As an ARRL section manager, I'm pretty much aware of who's doing what in my section (the state of Oklahoma). If you live within my section, you can call me and ask whom you should talk to about a particular activity. So barring any other source you may have (local radio clubs, for instance), I suggest that you contact your ARRL section manager. The managers' names, calls, addresses, and phone numbers are always listed on page 8 of the current issue of *QST*. and they will be most happy to help you find a mentor for any specialty you wish to learn about.

Finally, whatever your amateur radio interests, the VHF+ frequencies provide ways to explore them. This book will show you how.

Chapter 2

Designing Your VHF+ Station

Early operation on the VHF+ frequencies was quite a challenge. Before World War II almost all operation was experimental. Until the mid-1960s most VHF+ operation was the bailiwick of hams challenging the unknown frontiers of the bands. With the development of the FM repeater, operation on these bands flourished, and the emphasis shifted away from exploration and toward day-to-day communication. As a result, the VHF+ station is now a part of many amateur radio operators' shacks.

Before designing your own VHF+ station, you'll want to ask yourself several questions including: Where do I live? What are my amateur radio interests? What type of equipment (VHF base/mobile transceivers and handhelds) do I want to use in my station? Let's look at each of these questions.

Where do you live? Most of us, when we assemble our stations, must answer the question: "Where will I put it?" The ultimate answer to this question depends on where you live. Do you live in an apartment, a house with a small lot, a house with a large lot, in the country with virtually unlimited space, or, do you just want (or need) to have a station in your car? Are there zoning requirements regarding antennas or towers in your neighborhood? What safety standards does your community require you to meet when installing an amateur radio station? Be sure to check out the location of power lines and other obstacles to safe operation.

What are your amateur radio interests? Do you want to use CW? Do you want to upgrade right away or wait awhile? Are you a straight Technician class licensee, or did you also pass the Morse code test with your Technician class theory tests to earn your Technician Plus license? If you don't have HF privileges, do you want to stay on VHF, or do you plan to learn the code and upgrade so you can operate on HF?

Your interests are somewhat dictated by your current or future license class. However, even if you're a Novice or a straight Technician class licensee, there are many specialties you can pursue. These include weak signal VHF, packet, SSTV, ATV, and satellite. Others are traffic handling, public service, and equipment construction.

What type of equipment do you want for your station? The equipment you choose will have a lot to do with your answers to the previous questions. Depending on whether you want to get on HF or operate on VHF through the local repeaters, there are several equipment options for you to consider. If you want to purchase a handheld radio, you'll find there are several choices available.

Assembling a station for the VHF+ frequencies is similar to building an HF station, particularly for lower VHF+ frequencies such as 6 and 2 meters. However, for the higher VHF+ frequencies, with the exception of 70 cm, you'll probably have to consider building your own radio—either by designing it by yourself or by assembling a kit.

Six meters is the most popular VHF+ frequency, followed closely by 2 meters. There are single- and multiple-band transceivers and HF transceivers that include 6 meters. There are also several antenna suppliers for this band. If you're interested in 2 meters, you can find single and multiple band radios that include this band. Most manufacturers no longer supply equipment for 135 cm because of the previous uncertainty regarding the band. There are manufacturers who supply multiple band radios for 70 cm. Some of these units are designed for satellite work and some are also used for 2 meters. There are few commercially available radios for the 902 MHz band, or for any band above 2300 MHz. The principal equipment selection comes from a couple of manufacturers who supply kits.

Almost all of these commercially available radios are powered by 12 volts DC. Even most of the radios that accept power from commercial sources can also accept power from 12 volt DC sources. This means that these radios can be used mobile or in Rover operations. (See Chapter 17 for a description of a Rover.)

Here are some other factors to consider when choosing a radio:

• Most VHF+ transceivers are capable of running at least 10 watts output, which means that you might want to consider adding a linear amplifier to boost your signal.

• Almost all transceivers are capable of operating on FM. However, a few multi-mode transceivers do not include FM operation. Nevertheless, some transceivers not factory outfitted for FM operation offer it as an option.

• Most multi-mode transceivers contain built-in filters for more selective reception, or have provisions for adding them later.

• For CW, all transceivers have a built-in side tone, or feed back a sample of the signal through the audio section for you to hear what you are sending. Some transceivers even have built-in keyers.

• Most transceivers come equipped with a microphone. Although there is a headphone jack, you'll have to buy your own headphones. Some transceivers are shipped computer controllable; others are upgradable for computer control.

• Some transceivers include direct frequency entry, memories, multiple noise limiters, digital voice readout, digital recorders, digital signal processing, a built-in preamp, microphone processor, and ports for a separate receiver antenna and computer control operations.

Let's now examine in more detail some of the equipment you will need to consider for your station.

Base/Mobile Transceivers For VHF+ Frequencies

If you're a standard Technician class licensee, or you want to limit your activities to the VHF+ frequencies, you'll want to establish a station for those bands. Let's look at what makes up a base/mobile station for the VHF+ frequencies.

Your most important acquisition is your transceiver. As stated above, most major manufacturers offer units powered by 12 volts DC. This provides you with a radio versatile enough to use in your home with a DC power supply, or in your car with the car battery. As a result, you can take the same radio you use at home and hook it up in your car when you travel.

What should you look for in a radio? Two meters is by far the most popular of the VHF+ bands. Its popularity may be a deterrent to those new to amateur radio, but this needn't be the case. Just listen to the activity on your local repeater to learn the area etiquette (see Chapter 4 for more about this etiquette), and make your debut at a time when the repeater is relatively quiet. After you make your first few contacts, you'll wonder why you ever felt "mike shy."

If you live in an area where 2 meters is highly congested, or you plan to explore some of the other VHF+ bands at a later date, you might want to consider purchasing a dual-band, or even a tri-band, transceiver. The 70 and 23 cm bands are very popular, and interest is growing in both 135 cm and 6 meters.

Most manufacturers offer single-band transceivers. A majority also offer dual-band units. A few even offer "tri-banders." If you buy a tri-band radio, you can often choose your third band. I recommend that you ask either for the band most popular in your area, or for a frequency of particular interest to you. When I had the opportunity to choose the third band in my tri-band transceiver, I opted for 6 meters because I enjoy working DX on FM when the band occasionally is open via sporadic-E during the winter and summer months.

Transceiver manufacturers offer many interesting features that you may want to consider. For instance, you might want to look into obtaining a radio with a quick-release front control panel. This lets you remove the radio and take it with you when you leave your car, avoiding the possibility of theft. Such a panel also gives you the ability to mount the control panel on the dashboard—where you may have only limited installation space—and install the main portion of the radio in a roomier part of the car.

Some dual- or tri-band radios offer a repeater function. Almost all tri-banders have this capability. Note, however, that although these radios provide you with a crossband repeater option; they are not designed for the continuous duty that commercial-grade repeaters endure. Therefore, consider this as an option, but realize you can't use such a unit to replace a commercial-grade repeater.

Most transceivers now contain memories. There are two types of memories—one for frequencies and the other for phone numbers. While it's nice to have a number of memories, you'll seldom need more than 25 to 30 frequency memories or 5 to 10 phone number memories. It's important to be able to program at least one of your frequency memories for the occasional "odd split" repeater frequency offset. When looking at telephone memories, make sure the radio you choose lets you program enough digits to dial the repeater's access code and the telephone number (even a 10- or 11-digit long distance number).

Some radios will let you program a memory channel with a continuous tone-controlled squelch system, or CTCSS tone. This low frequency tone is used for access by some repeaters to reduce QRM. It's also used by some repeater owners to "close" the repeater to the general ham population. Most manufacturers offer this feature, either built-in or as an option. Without it, you may be "locked out" of some repeaters.

A wireless mike is another handy gadget that's available from a couple of manufacturers. If you've ever had your mike cable get tangled up in your steering wheel, you'll certainly understand the attraction of such an option.

How much power should your transceiver run? Twenty-five watts on 2 meters is more than adequate. Even 5 or 10 watts may be enough for most communications. However, it's nice to have a little more power available if you need it. Radios made for the higher VHF+ bands tend to have less power, because the price of the higher power components is a bit more than the manufacturer can pay and still make the radio affordable to most of us. Nevertheless, the lack of QRM on these bands and the difference in propagation can, most of the time, compensate for lower power.

The Handheld

The handheld radio makes it possible for you to take your ham radio station anywhere. These radios have come a long way since World War II, when they were more realistically "arm held." Today's compact radio fits inside your backpack, purse, or jacket pocket.

These ultra-compact radios aren't for everyone, however. If you're like me, and have to hold everything a bit farther away to see it these days, you may have trouble reading the lettering above the keys on the keypad. Also, the smaller the radio, the smaller the keypad. Pushing the right button on the smaller keypad may be a bit more difficult than on a bigger one.

Because you're carrying your entire station with you, you're also carrying your power source. That power source will eventually discharge, if you don't have a way of renewing it. Your best bet is to look for a handheld with a power-saving feature such as a battery saver function. This shuts off the receiver in the absence of a signal. An associated function is sometimes called the "sleep" mode. Sleep mode enables you to program the radio to turn off altogether after a certain length of time. The high/low power switch is a third power-saving feature.

When you purchase your handheld, consider acquiring a spare battery. Many radios come with a choice of battery. A bigger battery may provide longer times between charges, or it may enable you to run higher power. Evaluate your selections carefully before making a decision.

Look at how the battery is recharged. Exposed terminals can prove dangerous—even to the point of starting a fire. You wouldn't think that a lowly handheld radio could set your house on fire; however, that nearly happened to a ham in the midwest. He had scattered items from his school bag across his bed, and somehow the metal spiral of his notebook came in contact with both terminals of the battery of his handheld. In short order, the spiral was hot enough that the notebook and bedding started burning. If the ham hadn't noticed the fire, the results could have been disastrous.

Don't think this could happen to you? Some hams have reported burn holes in their pockets after filling them with their handhelds or spare battery packs and keys and other metal objects that have shorted the terminals!

How can this happen? Nickel-cadmium, or NiCd, batteries are designed with a low internal resistance. This allows them to deliver high current for many of their applications, such as for the handheld, power tools, and so on. If a nickel-cadmium battery pack has external contacts for charging purposes that are not protected either by design or diodes, or if a battery pack's terminals come in contact with a path that shorts it out, then a large current could flow for long enough to heat the battery and the object shorting it— possibly causing a fire or a nasty burn. Therefore, when using a handheld, be sure to protect the battery pack terminals from accidental contact with metal objects.

Features found in base/mobile radios are also nice options to have in a handheld. Check out a handheld's number and types of memories, CTCSS compatibility, and dual-band capability before you buy.

Because a handheld is a compact radio, you probably won't be carrying around a full-size Yagi. By design, and because of its compact size, the whip antenna, sometimes known as a rubber duckie because of its flexibility, is a good compromise. If you're having trouble accessing a particular repeater, however, you may need a bigger antenna such as a collapsible whip. There are several on the market designed for these little radios.

What about operating your handheld from the car? No problem! Most handhelds have an input port for a separate power source and also supply an optional cable that will plug into the cigarette lighter and radio.

If you use your handheld in your car, you'll need an external antenna, because the metal of the car is too great an obstacle for the radio signal to overcome. However, you'll need to watch for crosstalk when using your handheld with an outside antenna. To compensate for the small antenna supplied with the handheld, most manufacturers make the receiver very sensitive. Couple that sensitivity with a bigger antenna, especially one that has some gain, and you can receive unwanted signals from other nearby frequencies. There's nothing wrong with the radio, the antenna, or the other stations' signals. It's just one of those compromises you may have to live with.

For more detailed information about antennas, see the following section.

Antennas

Antennas for the VHF+ ham bands are commercially available up through 1296 MHz. Beyond that band, you'll want to consider building your own.

If you're going to use the radio at home, you might want to consider a base station antenna installation. If your radio includes more than one band, you'll need an antenna for each one. You can either purchase antennas or check the various amateur radio publications for construction ideas.

The most popular antennas are Yagis (beams) and quads. Yagis and quads are typically used on 6 meters through 23 cm. Both have multiple elements, starting with a minimum of three (two, for quads) on 6 meters and increasing to in excess of 25 on the microwave bands. On microwave bands, quads take the form of circular loops. Dish antennas are often used on 23 cm and above, particularly for the higher gain needed for EME work.

By convention, and with an eye to propagation considerations, vertically polarized antennas are preferred for FM operation. Several manufacturers offer dual-band verticals. Because of their light weight, they can be installed on the side of the roof, the side of a chim-

Ted Goldthorpe, WA4VCC, is shown here with antennas mounted on his Roving van during a recent ARRL VHF contest. (Photo courtesy KB4CSE)

ney, or on a properly guyed telescoping TV mast. If you have a dual-band antenna with a single connection and a radio with two ports, you'll need a duplexer to combine the signals of both bands into the same coax. These are available from some manufacturers. Make sure the duplexer you select is rated for the maximum power output of your radio.

If you've decided to install your radio in your car, consider how you feel about drilling a hole in the roof, trunk, or some other part of the car. This may not appeal to you if your car is new or nearly new. Even the owner of an older car may be sensitive about drilling holes that might affect the car's resale value. Fortunately, there is an alternative—the magnetic (mag) mount antenna. Unfortunately, there is a cosmetic drawback to mag mounts, too. Water can collect between the antenna and the car body. Consequently, you must pay careful attention to your mag mount. The combination of the vibration of the antenna on the surface of the metal coupled with any water that gets in between the metal and the magnet may cause rust. If this happens, you could be looking at a new paint job for the car.

Let's now look at antennas for each individual band.

For 6 meters, a 2-element quad or a 3-element Yagi is more than adequate for most operation. Even a dipole or a vertical antenna can be used with some success during good sporadic-E openings. However, if you want to work long-haul DX, you might need to add more elements to your antenna. Remember, though, the more elements, the more the gain and the narrower the beamwidth of your antenna. Because of the often rapidly changing aspects of sporadic-E propagation, signals may be coming to you from one direction one minute and from an entirely different direction the next.

If you're using a high-gain antenna, constant shifts in signal direction may leave you thinking the band has died when, in reality, the propagation has shifted. In order to keep up with the changes in propagation, you'll need to rotate your antenna frequently during a band opening.

For those considering the exotic mode of EME communication, multi-element and stacked array antennas are definitely necessary. Because much of the signal is lost on its travels between the Earth and the Moon, you'll want to direct as much of your signal as possible at that small target in the sky.

Two meter antennas are a bit more elaborate. Before investing in an antenna for this band, think about where you'll be concentrating your efforts. If you're principally planning to work meteor scatter, you'll want an antenna that has a rather wide pattern. Typically, the shorter boom (12 to 17 feet long) multi-element Yagis will satisfy this requirement.

If you want to work EME, you'll need a long-boom, higher gain antenna. Some exotic antennas have been built for this purpose. Paul Kelley, N1BUG, has built one such antenna array, which includes 24 four-element quads. With it, he has enjoyed much success "on the Moon."

For local FM or packet operation a vertical or three- or four-element Yagi will do.

Antennas for 135 cm are very similar to those for 2 meters. Multi-element antennas are available for this band. Should you wish to confine your contacts to terrestrial propagation, a single Yagi will suffice. However, if you plan to work EME, you'll want to consider stacking these antennas. Again, for local FM or packet, verticals or small Yagis are sufficient.

Yagi antennas for 70 cm contain many more elements for the same physical boom length as do 2 meter antennas. These antennas are, by design, of higher gain and narrower beamwidth than their 2 meter counterparts. For EME purposes, you'll need to stack more than one Yagi. Actually, this band is high enough in frequency to consider using a dish. For the FM user, verticals, particularly dual-band types, are preferred.

For 33 and 23 cm terrestrial communications, loop antennas are attractive options. However, for EME work, a dish is almost a must. For bands higher than 23 cm, the gain-versus-size issue comes down on the side of size. Therefore, many of those who experiment on these higher microwave frequencies tend to use a dish.

Vertical antennas are available for 23 cm. However, because very little FM operation takes place above 23 cm, few commercial vertical antennas are available.

Coax and Connectors

As for feedlines, above 2 meters most of the common coaxial cables—such as RG-8, RG-11, RG-58, and RG-59—present nearly unacceptable attenuation problems due to their construction. Therefore, you must look at more rigid types of coaxial cable, known as *hardline,* as a possible solution for feeding the antenna. Even for 6 and 2 meters, extremely low-loss coax cables are a must.

Connectors are also a very important part of the station and are sometimes considered the weakest link. Above 2 meters, the UHF type, more commonly known as PL-259 for the male connection and SO-239 for the female connection, are lossy because they are not a true 50 ohm impedance. While it is also not a true 50 ohm impedance, the type-N connector comes much closer to approximating one. Therefore, use this connector on all transmission lines above 2 meters.

Although it's important to use the right type of connector, it's *critical* to install the connector properly. Make sure the connection is mechanically and electrically sound. If it's not, an open connection or short could cause serious damage to your equipment, and may blow up your expensive linear.

Linears

Are you interested in a linear amplifier? The most common type is a power amplifier containing solid-state devices. These amplifiers, commonly known as "bricks" because they resemble a brick in size and weight, are available for power levels between 30 watts and 350 watts. If you need to work into a repeater that's at least 50 or more miles from you, or if you travel away from the normal coverage of your repeater when mobile, you might want to consider a brick.

Remember that you should *only* run the minimum power necessary to maintain the contact. That is an FCC rule. Also, running too much power into a local repeater can overload the front end of the repeater receiver, distorting your signal on the output.

For most weak signal work, a brick amplifier is more than adequate. However, you might want to consider a tube-type amplifier. There are also high-power tube-type amplifiers capable of running all the way up to the "legal limit" on these frequencies.

It's important to interject a warning here. While the power levels of brick amplifiers aren't very high, the RF they produce should still be considered dangerous. The RF produced by the high-powered tube-type linear should be considered *lethal*.

If you've purchased an amplifier and aren't sure of *any* aspect of its installation, get competent assistance. For that matter, if you're unsure of any aspect of installing any part of your ham station, get help!

Computers In The Station

How do computers fit in the VHF+ station? The use of computers in VHF+ work is growing, and software is available. Such software falls into two categories—logging and application. Logging software contains features to aid you in record keeping, tracking the number of grid locators worked, and contesting. The software application programs available include satellite, EME and meteor shower tracking, and grid locator calculations.

In Addition . . .

When designing your station, you might also like to keep in mind the growing interest in satellite communications and packet radio. Chapters 5 and 6 cover satellite communications and packet radio, respectively.

Chapter 3

Antenna Installation, Shack Construction, and Safety

Some of us take pride in having a shack that looks like something the Occupational Safety and Health Administration (OSHA) might want to condemn. However, there's a Chinese proverb concerning a disaster looking for a place to happen. Your messy shack may just be that place.

Let's take a look at how you might be able to improve conditions in your shack. I'll start from the top (the antenna installation) and work my way down to the station power source.

Your Antenna Installation

The antenna is an integral part of any ham station, and one of the most difficult (and dangerous) pieces of equipment to install. Before beginning any work on your antenna, outline every step. Draw a diagram and plot plan showing the location and height of the antenna support, the length of guy wires, the location of buildings, and the location and relative height of any power lines.

The most important thing to consider during an antenna installation is the location of any power lines. According to OSHA, the distance between an overhead power line and your antenna should be *no less than 10 feet*. Local ordinances may require even greater separation. Check with your electrical utility company and local government offices for the regulations in your area. Note the placement of the power lines and plan your antenna's location as if you expect it to come down. If your antenna is higher than the power lines, try to determine if it, the pole or tower, or the guy wires will fall into the lines on their way down. If you're planning to install your antenna on top of a dwelling with a service drop from the utility pole, make sure your antenna, pole, tower, and/or guy wires

won't fall on or anywhere near that service drop.

Next, choose your antenna support. Many VHF+ antennas are lightweight and can be safely mounted on push-up or telescoping masts. Many times these masts are installed on a roof. If you plan a roof-top installation, I recommend that you use a mast no taller than 30 feet. Although masts come in telescoping sections that reach heights of nearly 50 feet, the extra 20 feet makes the installation potentially unstable. The 27 or so feet of the mast, coupled with the usual 15 feet at the peak of the roof, places your antenna up around 40 feet. That's plenty high enough for most casual weak signal work. In fact, Mark Ammann, KMØA, has worked more than 300 grid locators on 2 meters with an antenna installation at a height of less than 30 feet and 150 watts!

Always use an assistant when installing antennas. Any antenna installation will be easier and safer if you have help. Make sure you and your assistant wear hard hats, just in case any of the antenna hardware finds your head a good target. Always use an electrically safe ladder—one made from wood or a nonconducting material—for any antenna work. Your assistant should hold the ladder steady the entire time you're on it.

If you plan to install a telescoping mast on your roof, make sure it's safe and dry. If the roof needs repair, stay off until it's been fixed. If it's wet, go and fill out QSL cards instead. When you go up on the roof, take a handheld with you. While it may be a bit embarrassing to ask someone on the repeater to come over and reset your ground-to-roof ladder should it fall, it's far safer to call for help than to jump off the roof and reset the ladder yourself. Remember, you can fall off a roof. Know where cables, guy wires, and tools are in relation to your body. Wear shoes with the

Two of the inside cross booms of the EME array of Bob Taylor, WB5LBT, sustained damage from Hurricane Andrew. When repairing damaged equipment be sure to observe all safety precautions.

necessary traction for walking on the incline of the roof. If the incline is too steep for you, stay off.

When installing the mast on the roof, make sure the bottom of the mast has no way to move. Use a tripod that's securely bolted to the roof and reinforced from underneath. Before you mount the rotator and the antenna on the mast, calculate the amount of guy wire needed to reach the top section of the fully extended mast. (Do this beforehand, as a part of your project diagram.)

Next, carefully extend the mast, without the rotator or antenna installed, and attach the guy wires to the guy points securely. Leave enough slack so the mast doesn't bind when it's fully extended. Now, install the rotator first, then the antenna. *Do not* attempt to install the rotator and antenna simultaneously. Never install an antenna that's too large for the mast capacity. If you have doubts, go to a store that sells television antennas and look at the largest log periodic they sell. If your antenna is bigger than that, use a tower or some other support. That long boom antenna may seem relatively light when you lift it; however, it presents a tremendous torque when whipped back and forth by 60 mph winds. As a matter of fact, there's enough torque to damage the rotator break and cause the antenna to freewheel on the mast, or even twist the mast. If you want to install a long boom antenna on the roof, use a properly installed and guyed roof-mounted tower, and a mast with a minimum outside diameter of 2 inches. Always use a rotator that can support the wind load of the antenna. And always use guy wires that can handle

the wind load of the antenna, rotator, and mast. Make no compromises on anything, *ever.*

Once the antenna is in the air, make your mast turning adjustments for calibrating the rotator control box. (I've found that a large pipe wrench snugly affixed to the base of the mast makes this job very easy.) Then, tighten the guy wires and the bolts on the tripod to secure the mast.

Your antenna/mast installation is now a lightning rod. To counteract this characteristic, attach one end of a no. 8 wire to the base of the mast and connect the other end to an earth ground via the shortest path. An earth ground is an 8 foot long copper rod with its entire length pounded into the ground. I know that after six feet, the hammer is heavy and the arms are real tired; however, you must resist the temptation to saw off that remaining length above ground. There are no substitutes for a proper earth ground.

If you're planning to put up a tower, check first with the local officials concerning zoning regulations, engineering requirements, and restrictions regarding tower installation to see what's permitted in your area. The FCC's regulation PRB-1 does supersede local restrictions, but only to the extent that the local regulations should make "reasonable accommodations" for your antenna installations. It's up to you to find out what those "reasonable accommodations" are.

After determining what you can erect, consult with tower company engineers concerning anything that might be required for the safe installation of the equipment you've selected. Ask about the amount of concrete required for the base, the type of guy wires to use, the requirements for concrete at the guy anchor points, and the various additional pieces of hardware needed.

Always use a tower and rotator strong enough to withstand the wind load of the antennas you're installing. If your tower is to be guyed, make sure the guy wires are the proper strength and won't travel into power lines should they snap. Carefully follow the manufacturer's instructions when installing the tower. Also, consult *The ARRL Antenna Book* as a source of background information while installing your tower/antennas. The first chapter covers safety procedures; read it before proceeding with any work.

It's of paramount importance that you *never* install used antenna support equipment. Stay away from used guy wires, masts, and/or towers. If any item is used and/or damaged in any way, reject it. I would go so far as to caution you about putting up a used antenna. It's far better to pay the extra money for new

Marshall Goldblatt, W4EMB, and Pete Hein, K1FJM, are examining the remains of Marshall's M² 2 meter antenna array, which was demolished by a recent hurricane. Whether you are erecting a new antenna array or repairing an old one, always remember to use only the best components and to observe safety precautions. (Photo courtesy W4EMB)

equipment, than to risk injury with used equipment that may not be in top condition.

When performing any antenna installation, give yourself plenty of time to complete your task. Take your time putting up the antenna and keep your mind on your work. It's not a good idea to put up any antenna hours before you absolutely must use it. I know that Field Day encourages swift emergency installation of entire stations; however, I think it would be prudent to go through a dry run exercise before the quick installation of any antenna setup. This way, you can proceed with care and work out any bugs associated with your particular installation.

If you plan to work when it's hot out, have plenty of thirst quenchers, juice, and bananas available. Take frequent rest breaks and replace your lost liquids and potassium with these items. Sodas and beer are not good sources of the nutrients your body needs when performing hard physical labor.

When on the tower, wear a safety belt designed to handle your weight. Check the belt every time you plan to use it for cuts or nicks in the leather. If it's damaged, discard it and get a new one. As with a roof-top installation, have all members of your team wear hard hats. You may forget about the antenna or gin pole that's just above you, as you're climbing up the tower. Besides, you never know when a wrench or an antenna may come down. Resist the temptation to follow the item down. If something falls, even if it's your new $300 antenna, let it go. It can be replaced; you can't!

Ground your tower to at least three 8 foot ground rods installed in a triangle around the base of the tower. These ground rods must be connected to each other, and to each leg of the tower. If your tower is a crank-up model, be sure that someone you know, who can be contacted in an emergency, knows how to lower it.

One final note about antennas: Don't work around or near any antenna that has power applied to it. According to the American National Standards Institute (ANSI), your body has an increased susceptibility to RF damage in the 30 to 300 MHz range. The 6 meter, 2 meter, and 135 cm ham bands all fall within this frequency range. When thinking about RF energy, keep in mind that your microwave oven uses microwaves to cook meat. Like it or not, your body is a piece of meat. A friend of mine once worked on a fire control radar antenna when it was live. He told me later that he felt parts of his body get warm from the RF radiation. The jury is still out on whether or not RF radiation can cause cancer or cataracts, but it's better to be safe than sorry.

Your Shack

The electricity to your shack should be on separate circuit breakers, and responsible members of your family should know where they are located. Make sure the breakers are the kind that can be "locked out." Lock-out circuit breakers are designed so locks can be attached to them. This prevents a breaker from being thrown "on" when it's turned off. The wiring to the shack should be the modern three-wire (hot, neutral, and ground) type, and of adequate current carrying capacity relative to the circuit breaker used. Make sure the wiring and circuit breaker installation is done by a qualified electrician. All receptacles should be the three-wire grounded type. Make sure there are plenty of receptacles available, so no single receptacle is loaded with a number of plugs.

The shack ground should be attached to an 8 foot ground rod located as close to the shack as possible. Take care to protect the shack from electromagnetic pulses (EMP) and lightning. Install surge suppressors in the electrical lines to prevent surges induced by power company variations and nearby lightning strikes from damaging your equipment. Also install lightning arrestors in the coax lines. Provide a way to ground all coax lines during a lightning storm. Make sure the ground wire connecting each piece of equipment is flat braided copper, at least a half inch wide, in order to prevent the ground wire from becoming an antenna. Each connection to every piece of equipment should be mechanically sound, and should be checked periodically to maintain that connection.

When working around semiconductor devices, provide protection from electro-static discharge (ESD). You can do this by using a wrist strap attached to a static-resistant mat and chains attached to the bottom of your chair.

Make sure your work area is properly lighted. Also make sure that the work bench and/or operating table is sturdy and uncluttered (I know that's hard to do sometimes). Check to see that the desk and chair heights are compatible and that the chair is ergonomi-cally designed, supporting your weight and back. You can do long-term damage to your back by working in an area that has the wrong height relationship between the desk and chair, or by using the wrong type of chair.

Your Conduct

Unless you are qualified, never work on anything "live." By qualified, I mean that you have attended a school, have received proper training, and have been certified to work on the equipment. Two things should make you think twice about working on "live" equipment. First, voltages as low as 40 volts can kill you, and 3,000 volts can overkill you. Therefore, in order to work on high-voltage equipment, you must have special high-voltage certification. Second, leaking RF may cause some physical problems. An RF burn is most painful and heals very slowly because you cook from the inside out. Ask that hamburger you cooked in the microwave oven how long it will take to recover from its "RF burn."

Also, as the following story illustrates, stay off the air during electrical storms. No band opening is that important! Finally, remember the safety rule: "When in doubt, don't."

A Safety Story

A graphic example of what can occur when lightning strikes happened to Pat Stein, N8BRA, in the fall of 1991. Pat had a run-in with lightning that was quite spectacular. Around 5 a.m. a bolt of lightning hit his electrical power company service drop from the

Dan Burns, N4YKD, and Ray Veldran, N4KWX, are shown here raising their antennas in 40 mph winds during an ARRL VHF contest. When doing this kind of antenna work, or any such work for that matter, remember to observe all safety precautions. (Photo courtesy N4KWX)

power pole outside his house. The energy traveled down the service drop into the service panel inside the house. Pat reported that circuit breakers flew out of the service panel across the room. He said it appeared that fireballs went everywhere around him and his family within the house.

The effect to the outside air was instant ionization. Pat has a tall tower that is well guyed and grounded. Normally, that type of installation provides a bit of a zone of protection. However, when the air was ionized from the lightning strike, the inevitable happened.

The second strike occurred about a minute and a half later. This bolt hit the tower. It damaged all his antennas on the tower, and caused the Cushcraft 3219 two meter Yagi antenna to explode. (He and his neighbors later found pieces of the elements over 500 feet from the tower.)

The bolt continued to travel down the tower, including down the guy wires, the antenna rotator cable, and the hardline, and on into the house. The hardline was ripped open like a banana peel, and the cable was blown off the rotator control box inside the shack.

In the aftermath of the two strikes, they found that almost every light bulb had been blown out of its socket. Every appliance except the electric stove, the washing machine, and the toaster was damaged to some degree. Pat said that about half of the mother board in one of the televisions was charred beyond recognition, every piece of ham radio equipment was damaged, the pump in the water well located some distance from the house was damaged, and much of the interior electrical wiring was destroyed.

Additionally, Pat found that there was a 6 inch trench dug in the ground surrounding the tower where the dirt was blown away. The dirt was also exploded away from the concrete blocks holding the earth anchors. He stated that a neighbor who saw the second strike thought the whole house had gone up in flames when the area around the house lit up due to the ionization.

When Pat dug up his hardline, he found that it had arc damage. The lightning had arced to the underground air-conditioning ducting several feet from the hardline, and to the house wiring in the attic.

Pat feels very fortunate that the house did not go up in flames. He attributes that good fortune (if there can be any in this disaster) to the type of insulation he used in the walls and attic. He stated that it had a very high flash point and provided a measure of protection from the red-hot wiring when the wiring was receiving the extra energy from the bolts. Pat said that although it represented some terrifying moments for his family, incredibly, two of his children slept through the entire affair.

While you can make every effort to protect yourself from the nearby hits, you cannot afford yourself 100 percent protection. Be aware of what lightning can do and respect it. I am sure that Pat has a new appreciation of the power of lightning. And after reading this story, I hope you do, as well.

A Final Word About Safety

While I've by no means entirely covered the subject of safety, I hope the preceding information will make you think about what you're doing. The hardest subject for me to write about is the Silent Key. And the last thing I want to write about you is that you've become one.

Chapter 4

Repeater and FM Simplex Operation

For many new operators, their very first introduction to ham radio is via the FM repeater. However, repeater operation on the VHF+ ham bands actually began in the early 1930s. One of the first stations was relay station W1AWW, which operated on the old 5 meter band from a lookout tower near Springfield, Massachusetts beginning around 1932.

While it was a bit of a historical accident, this early repeater operation was on FM! Because the rig was a modulated oscillator, there were more FM than AM products on the signal. However, in those days it didn't really matter.

Owing to the theory "If you can't see it, you can't work it,"* many felt that communications beyond line-of-sight on the VHF+ frequencies weren't possible. Therefore, construction of relay stations on the Atlantic seaboard helped boost signals of small transmitters on the old 5 meter band enough to get past that pseudo barrier.

In the beginning these "repeater" stations were only set up on weekends, when hams had the time to play. Stations were manned all the time they were set up because, in those days, the relay station consisted of a transmitter and several receivers. When the relay operator heard a station on a particular receiver, he would patch the audio from that receiver to the transmitter. After the station ceased transmitting, the relay operator would disconnect the audio patch and wait for the next station to start transmitting. If it was transmitting on another frequency, the relay operator

Although wryly attributed to Ed Tilton, W1HDQ, the origination of this quote has been lost in history. If Ed did originate it, and was serious (and only he knew, but has since forgotten), then he was later proven wrong many times— most notably by his own actions, when on November 24, 1946 Ed worked G6DH on 6 meters for the first ever trans-Atlantic VHF QSO.

would have to connect that receiver's audio to the transmitter. As you can see, the process was cumbersome, at best!

The success of these stations brought the same problems we experience today. With more interest came more operators—and overcrowding. Additionally, operators who felt that over-the-horizon communications were possible on the VHF+ frequencies looked with disdain at the use of relay stations for communications. Championed by Ross Hull, their cause was given much impetus when in 1934 his experiments in tropospheric enhancement were successful.[1]

As more and more operators constructed higher power stations to use this form of propagation, operators using less powerful equipment were either obliterated by the stronger signals or squeezed out of the band.

Interestingly enough, owing to Hull's discovery, the second great dilemma of the VHF+ ham bands was born. This dilemma is frequency management, or what is more commonly known as band planning. It's a problem that, except for some "band-aid" type solutions along the way, continues to this day.

While World War II shut down the ham bands, it didn't curtail interest in FM operation. Given a boost just before the war by Edwin Armstrong, commercial users, particularly point-to-point communicators, found FM superior to AM for communicating. Following the war, hams took a cautious but renewed interest in the FM mode.

Even though narrow-band FM experiments on 75 meters in 1946 showed that FM was superior to AM, hams were reluctant to embrace it. Unfortunately, the almost parallel development of SSB all but doomed FM operation on the HF bands.

However, the VHF+ ham bands were another issue. Because of the growth in commercial FM operation, the FCC required all but the broadcast users *and hams* to convert from wide band FM (15 kHz) to narrow band FM. This action immediately created a surplus market for radios that were no longer usable in commercial service. Nevertheless, hams were still slow to embrace this form of communications. It took the availability of equipment, the demonstrated superiority of FM over AM in overriding manmade noise, and the knowledge that the mode was less apt to cause audio rectification problems in non-amateur radio equipment (in those days your neighbor's new Hi-Fi) to cause hams to seriously consider FM as a viable mode. Couple all these factors with the growth of the ham population on the VHF+ frequencies as a result of the newly created Technician and Novice class licenses (in those days Novices had voice privileges on the 2 meter ham band between 145 and 147 MHz, the same as the Technicians at the time), and suddenly there was interest in calling attention to FM repeater operation.

Despite all the factors pushing toward FM operation, there was continued resistance from AM operators. In the 1950s, early repeaters were AM modulated. As a newly licensed ham in 1960, I watched the transition from AM to FM. Initially, hams in the southern California area constructed and set up AM repeaters. However, as FM modulation continued to be demonstrated as the superior mode because of its being impervious to heterodyning and fluttering, more and more repeaters were built for FM. Also, because the surplus of FM equipment included repeaters, more repeater owners went with FM. Before long, FM repeaters were the dominant form of relayed communications.

Parallel to this development, commercial manufacturers entered the market with compact and easily transportable solid-state radios. This lead to another interesting turn of events.

In the early 1960s, many hams who had General class privileges and operated mobile confined their operation to 75 meters for point-to-point communications. However, the problems of skip and signal loss stemming from locations in valleys or under bridges plagued these operators. Couple these problems with the enormous antennas required, and the fact that one had to have the higher class license to operate on 75 meters, and you can see that this type of AM operation had serious drawbacks. Therefore, as word spread about the superiority of FM VHF communications and the availability of more and reliable repeaters, HF mobile operation for close-in communications all but ceased. Suddenly, more hams than ever were looking

to the repeater for reliable local communications. Even the die-hard 6 meter operators began to flee that band for the 2 meter FM repeater.

As a result, the VHF+ frequencies became increasingly crowded. The FCC repealed Novice phone privileges on 2 meters partly because of increasing pressure on that band to accommodate all the repeaters. At the same time, the FCC gave the Technicians use of all but the bottom 100 kHz of the band in order to accommodate this growth and to "keep them legal," because by now repeaters were being activated in the "General" (147 MHz) portion of the band.

During the late 1960s and early 1970s repeater growth mushroomed, so much so that bands began to fill up. Repeater operation spread first to 70 cm and then to 135 cm. Eventually, the 23 cm ham band began filling up with repeaters. With all this growth, repeater wars developed. The challenge as to "who owned the frequency" led to the formation of repeater coordinating councils.

While this growth continued, other interest groups felt the pressures of these operations. It was the conflict of 1934 in reverse. Now, however, the pressure was on the weak signal and satellite users.

These developments led to band planning. With input from representatives of the various users of the bands, the ARRL developed band plans to accommodate each type of user in his or her portion of the band. While these were gentlemen's agreements, it was hoped that the spirit of cooperation, self-policing, and self-discipline would hold the line between each user type. This has worked in many cases. Recently, however, weak signal users felt that after the loss of the bottom 2 MHz of the 135 cm ham band (which included their segment of operation), they needed protection from encroachment by the tremendous growth of repeaters on that band. Ultimately, the FCC agreed and developed regulations that assigned a portion of the band to "experimental" (weak signal) use, exclusively.

Where are we today? It's almost automatic. You get your license, you go to the ham store (or hamfest, or telephone), purchase your handheld radio, open the box, connect everything, and start transmitting. It's almost that simple. However, there are a few technical aspects of operating a repeater that you should know. You also need to learn FM repeater operating protocol.

Let's start with some general knowledge. Nowadays, most repeaters are located on tall buildings, high hills, tall towers, or anything else that's tall compared to the surrounding terrain. In Oklahoma City, for example, many repeaters share space on TV towers

Whenever I go somewhere and expect to rent a car, I pack my 2 meter radio and my SQLOOP antenna. I use the antenna by mounting it on the back of my rental car, as shown in this photo. Installation simply requires some rope, an old mag mount, a piece of continuous threaded stock, a couple of nuts and washers (for positioning the antenna on the stock), a PL259 coax connector (that screws onto one end of the continuous stock and inserts into the SO239 connector on the mag mount), and a length of coax. All items, including the radio, easily fit into my briefcase or carry-on bag.

because they're the tallest structures around. In areas that have mountains, repeaters share space on the top.

When you use your handheld, you'll notice that you're operating a radio that is, to use the commercial term, "channelized." Everyone using a particular repeater transmits on the same frequency. This is called the "input frequency," because the repeater receiver listens on that frequency. Everyone also listens on the same frequency, called the "output frequency," because the repeater transmits on that frequency. These two frequencies are separated by a certain number of kilohertz (on 6 and 2 meters) or megahertz on the other bands. The separation on 2 meters is 600 kHz. On 6 meters, the separation is 500 kHz to 1 MHz. On 135 cm, the separation is 1.6 MHz. On 70 cm, the separation is 5 MHz. And, on 33 and 23 cm, the separation is 12 MHz. All these separations are referred to as "standard splits." In rare circumstances, a repeater is set up to operate on something other than these pairs of frequencies; that set of frequencies is called a "non-standard split."

By consensus, through the band plans mentioned above and under FCC regulation, repeater operation is conducted in certain portions of the bands. See the Appendix for the portions of each band that designate repeater operation. By regional agreement, a repeater is assigned a "frequency pair" by the local or area coordinating council. This frequency pair sets the input and the output frequencies for the repeater. To be assigned a frequency pair, you must apply to the

repeater coordinating council for your area. To find the name of your local council, check the *ARRL Repeater Directory.*

The input frequency is also known as the "offset frequency" because it's offset from the transmit frequency. Depending on where in the band the frequency pair is located, the offset frequency is either up, plus (+), or down, minus (–), from the transmit frequency. For example, if the repeater is assigned an output frequency below 147.000 MHz, the offset is usually minus and visa versa.

The repeater is equipped with an identifier that transmits the callsign of the control operator. This identification, by law, must take place at least every ten minutes and at the cessation of activity on the repeater. It can be voice or CW at a rate of around 13 to 16 wpm.

Gone are the days of the operator-switched receivers. Operator switching has been replaced on some repeaters by an automatic function called "voting." Voting works by using several receivers at various sites around the area to be covered. These receivers are linked "crossband" or via dedicated phone lines to the transmitter site, which may also have a receiver. The receiver hearing the loudest signal is electronically "plugged in" to the transmitter by the voting system, and that signal is then retransmitted.

This physical separation helps alleviate a problem called "desensing"—something that happens to a receiver when a strong signal is in close proximity.

During desensing, the receiver front end is so overwhelmed by this strong signal, even if it's not on the same frequency, that it can't hear anything else. This could be a problem when the transmitter of the repeater's signal sounds so loud to the receiver that it can't hear anything but the transmitter. All distant, or weaker, signals are blocked out.

For the repeaters with on-site receivers, desensing is combatted (although not totally solved) using a "duplexer." This duplexer isolates the transmit signal from the receiver. Consequently, instead of overloading the receiver, the signal is several tens of dB weaker, or down from, its normal signal strength and is not nearly as much of a threat as it originally was to the on-site receiver.

Sometimes repeater receivers are subject to another type of interference—that from users of another repeater operating on the same frequencies. Occasionally, conditions are such that signals from another repeater on the same frequency are so loud that they render the first repeater inoperable. It's possible to prevent this kind of interference by using a continuous tone-controlled squelch system, or CTCSS tones. These tones operate on low frequencies and low levels that aren't normally passed through the repeater to the transmitter. A repeater receiver that's set up to operate with CTCSS has its input tuned to "hear" the correct tone. If it doesn't hear the tone, the receiver doesn't pass the signal on to the transmitter for retransmission.

While CTCSS tone access is designed to keep unwanted signals from inadvertently keying the repeater, it can also be used to deny access to those you don't want using the repeater. Although this practice isn't illegal, it is looked upon with a bit of disfavor because by law all amateur radio frequencies are open for use by all authorized (by class of license) amateur radio operators.

The "autopatch" is another device that's connected to and works through the repeater. An autopatch is a phone patch that automatically connects the repeater to local phone lines. Via the autopatch, an amateur can make a phone call to a friend or family member, or even contact emergency services to report an accident or emergency. While the autopatch is convenient, it can be misused if operators are careless about the type of communications they make over the patch and the length of time they use it.

While the FCC has made it legal to "order a pizza" over the autopatch, it's still illegal to use the patch to conduct day-to-day business activities. It is up to the repeater user to police himself as he determines the uses to which he will put the autopatch.

Each repeater is managed by a control operator. This person is responsible for the legal operation of the repeater. Therefore, it's important that you do nothing that would be considered illegal. Your illegal activities would not only jeopardize your license, but could imperil the license of the control operator.

The length of time a single operator uses the autopatch is generally regulated by a timer. Many timers are set to cut off the autopatch after three minutes during normal usage times. Some are designed to cut off after only a minute during peak times, such as during daily commuting hours. Most of the time an electronic verbal warning is given in such a way that both parties hear it approximately 30 seconds before the patch disconnects.

Most autopatches are "open"; that is, any ham can use them. However, some repeaters may restrict the use of the autopatch to certain members of a club or informal group. If this is the case, the autopatch is known as "closed" and certain access codes, known to the group, must be entered before dialing your telephone number in order to use the autopatch.

Even "open" machines may use the "*" and "#" buttons as controls to keep someone from accidentally initiating a telephone call. The phrase "star up and pound down" means that you must use the "*" button ahead of the phone number to access the autopatch and the "#" button after the phone call is completed in order to "hang up" the phone.

A timer is also used to keep one operator from monopolizing the repeater. Again, it is usually set for three minutes during "off peak" times, and for only 30 seconds during peak times. How do you know if you "timed out"? You know if you let up on the mike and hear an electronic voice say "You timed out!" It's a bit embarrassing to have this happen, so watch your time.

Another device used with the repeater is a "courtesy tone." You'll hear the courtesy tone at the end of a repeater transmission, when whoever is talking lets up on the PTT button. It's called a courtesy tone because it creates a pause of a second or so to allow someone who isn't part of your QSO to break in and say something. Perhaps the person has an emergency to report, or maybe he just has something to add to the conversation. Once the tone has sounded, the original parties are free to continue their QSO.

Even after the tone has sounded, it's a good idea to pause a second or two longer. If the repeater is equipped with a timer, the tone also indicates that the timer has been reset. If you start transmitting before the tone sounds, you'll pick up on the time remaining from the previous transmission. This means you could time out fairly quickly during your transmission.

What about protocol? Unlike HF QSOs, you don't "call CQ." Instead, you simply key your radio and announce, for example, "WB5LUA monitoring (or listening)." If someone else, let's say AA5C, is listening and wants to talk to you, then he will say, "WB5LUA this is AA5C" and listen for your reply. In such a fashion, a QSO begins.

Incidentally, jargon you may be used to from CB, police, or HF ham radio operation is generally not acceptable on the repeater. Instead of saying, "My QTH is . . ." simply say, "I'm located at . . ." And instead of saying "My handle is John," say, "My name is John." The term *handle*, by the way, was popular when I first got into ham radio, but it has since fallen into disuse and subsequent disfavor.

Emergency communications *always* have priority. While it varies from repeater to repeater, the typical protocol is to announce your emergency, such as "This is WA6PDE and I am at the scene of an automobile accident. Can anyone assist me by calling the police?" However, if you are too excited to dial the number or you do not know the access codes, ask for assistance. Chances are very good that the control operator or someone else will come to your aid and help you pass your emergency communications.

Above all, however, be calm and have your thoughts collected so that you can clearly and correctly give the information to the operator. When talking to emergency personnel, give detailed information about the situation. It is not necessary to explain autopatch to the operator. They are used to receiving emergency calls from every type of communications.

If you are using a "911" service, however, at the beginning of the conversation clearly identify yourself as being at the site of an emergency, passing emergency traffic. Most "911" operators are trained to recognize the ham radio operator and the cellular telephone user so they know the difference between a call from a person who has come upon the scene of an emergency (such as you, the ham radio operator) and one from a person involved in an emergency, such as in a home.

There is another important reason for letting the operator know that you are at the scene of an emergency. If you get disconnected, if the "911" service is using the enhanced feature that identifies the physical location of the phone number, the operator will dis-

patch emergency personnel to the location identified in his records. It will do no one any good to have an ambulance dispatched to the base of a radio tower when the emergency is clear across town.

If a QSO is in progress, then after you hear someone stop talking (and before the courtesy tone sounds) announce your emergency or simply say the word "Break." On this last point, however, use of the word "Break" has different meanings to different repeater users. To some it means you simply want to break into their conversation. To others it means you have emergency communications. Find out from the repeater control operator what the protocol is for that repeater.

Nevertheless, the use of the double words "Break-Break" has pretty much been accepted as meaning that you'll be following your transmission with emergency communications.

After you have reported the emergency, if you can, stay around until the emergency personnel arrive on the scene in order to further direct them. However, do not become involved in the emergency. Do not render assistance unless it is a life or death situation and you are confident of what you are doing.

For more information on training for and handling of emergency communications, consult your local Emergency Coordinator. If you do not know who he is, contact your ARRL section manager.

Whenever you use the autopatch, whether in an emergency, or for routine calls, you must identify your station as shown in the following example: "This is K5CPZ on the patch." Once you've completed your call, you must identify your station again: "This is K5CPZ clearing the patch." Again, check with the control operator for the required protocol.

One other word about "identifying." It is particularly annoying to listen to someone "kerchunk" or key up a repeater without identifying himself. To some operators, this may indicate that someone has emergency communications and is trying to, but is not quite able to, make it into the repeater. The potential ramification of kerchunking is that someone may spend unnecessary time listening for another operator just because that person didn't have the courtesy to identify himself. By law, you must identify your initial transmissions anyway, so don't leave somebody else hanging.

As I said, using a repeater will probably be your first activity as a newly licensed ham. There should be nothing intimidating about it. The normal, day-to-day users are usually friendly and want to help you "learn the ropes." Don't be afraid to pick up the mike (or handheld), key it, and join in. And don't be afraid to participate in emergency communications. The following story, used with permission from the March 1994 edition of the "Section Leader" newsletter, illustrates just how you can do so effectively and possibly even save a life.

Ham Radio Helps Save A Life

When Scott Montgomery, N9GLL, left for work that December 16 morning, he didn't expect anything other than the usual heavy traffic. But on passing a shopping mall, he noticed a small car parked with its hazard lights blinking continuously. Its door was wide open and an elderly man in the driver's seat was leaning back in an awkward position.

After Scott had driven on a short distance, he suddenly decided that what he had seen made no sense. He returned to the small car and rolling his window down, asked the elderly driver if he was all right.

The driver apparently couldn't talk, but gestured repeatedly toward his chest, which Scott (it turned out correctly) interpreted that the man was having difficulty in breathing, which is one sign of a possible heart attack. Additionally, he didn't look well at all. Scott had a 2 meter radio aboard, so he radioed his dad, Jack Montgomery, K9DQU, with "Priority Traffic."

On connecting with his dad via the "Big Mac" 2 meter repeater (operated by the Metro ARC), he advised that an ambulance was needed and quickly.

Jack called 911 requesting an ambulance and was transferred to the Chicago Fire Department (CFD), the usual procedure. With Jack talking to Scott on 2 meters, and at the same time talking to the Fire Department via 911, the need for the ambulance and as much detail as was available was passed on to the CFD.

Scott stayed at the scene to pin-point the ambulance's destination. Meanwhile, Jack monitored the 2 meter radio channel to keep it open, should further emergency assistance be required.

When the CFD paramedics arrived, they went to work on the sick man immediately, indicating that the situation was indeed serious. The ambulance driver then radioed for a Fire Department engine company to help with needed additional manpower. And at his request, Scott watched for and waved the engine company into position at the scene. Eventually, the engine company radioed the Chicago police to secure the sick man's vehicle, since obviously he was in no position to drive at all.

While all this was going on, Scott assisted the paramedics in putting the man aboard the ambulance, which took off for the hospital with the paramedics still working on the patient during the transit.

Before Scott left the scene, both the fire engine company personnel and the CFD paramedics thanked and praised Scott profusely for his assistance to a citizen in need, for there seemed no reason to doubt that Scott might have saved the man's life.

Linked Repeaters

Probably the best example of a system of linked repeaters is the "Zia Connection." This system, which covers most of Arizona and New Mexico, as well as certain western areas of Texas, southern areas of Colorado, and eastern portions of southern California, consists of 13 linked repeaters on the 2 meter band.

The system is run by Milt Jenson, N5IA, who, along with five others, operates and maintains these repeaters night and day. The repeaters provide linked communications for both regular users and motorists passing through the southwest. For more information, contact Milt at Rt. 1, Box 176, Duncan, Arizona 85534.

FM Simplex Operation and Hidden Transmitter Hunting

While repeaters seem to get the glory, there are other forms of communications on the VHF+ frequencies. Simplex operation takes place on the simplex frequencies without much notice from repeater users. So, if you're a bit bored with the same old group of people on the repeaters, tune to the simplex frequencies. You'll find a whole group of new friends.

Hidden transmitter hunting is another FM simplex activity. When I was in high school, most hidden transmitter hunting took place on 10 meters. However, today most takes place on 2 meters on an unused simplex frequency.

The object of hidden transmitter hunting is for one person, called the "fox," to transmit a signal on the 2 meter simplex frequency from a secret location. The "hunters," using direction-finding antennas and their own ingenuity, attempt to find the fox. Usually, the first hunter to track down the fox serves as the fox for the next hidden transmitter hunt.

In addition to being lots of fun, hidden transmitter hunting also serves a training purpose. Every once in awhile someone suffers the embarrassment of jamming his or her mike between the seats of the automobile, thereby keying the mobile radio. Feelings get a bit frayed if the input of the local repeater is blocked by an unknown signal for any length of time.

Enter the hidden transmitter hunter, complete with his sophisticated antennas and receivers. With the work of three or more of these highly trained amateurs (trained by winning so many hidden transmitter hunts, of course), triangulation and subsequent loca-

Hidden transmitter hunting is one of the fun activities on the VHF+ frequencies. John Brassfield, N5SAM, is ready to go find the "fox" with this three-element quad strapped to the car's passenger door.

tion of the offending signal is relatively easy. All that's left to do is to good naturedly embarrass the offender at the next radio club meeting.

In addition to being fun and practical, there's a serious side to hidden transmitter hunting. Hidden transmitter hunters have also participated in locating illegal activity.

Have Fun

In this chapter we've looked at the most popular use of the VHF+ frequencies, that of FM operation. There are many ways to use your FM radio: on a repeater, with your friends in a round-table discussion on simplex, or to find that elusive fox on a hidden transmitter hunt.

Reference

1. Ross A. Hull, "Air-Mass Conditions and the Bending of Ultra-High Frequency Waves," *QST*, June 1935, pp. 13–18, 74, and 76.

Chapter 5

Satellite Operation

As a high school student, I can remember bursting with pride when I was asked by my fellow students about the ham radio satellite orbiting the Earth. Although I didn't know much about amateur satellite communications at the time, I knew that hams were making a name for themselves with the successful launch of the first OSCAR (Orbital Satellite Carrying Amateur Radio) satellite.

Today, hams are maintaining that reputation with successful launches of more than three dozen satellites. Nearly twenty satellites that permit voice and packet communications are presently in orbit.

In recent years, satellite communications have been joined by contacts with astronauts. Both the United States, through the Shuttle Amateur Radio Experiment (SAREX) program, and Russia, through the MIR space station, have given hams on Earth the opportunity to communicate via voice, packet, and television with hams in space.

Amateur Satellite Communications

Communicating via satellite is kind of like communicating via a repeater. I say "kind of" because, while there are many similarities, there are also many differences. First, the satellite is always on the move. This constant rotation pretty much necessitates that you have a computer, so you can use the software that predicts the current whereabouts of the satellite you want to work through. Second, the satellite operates crossband. Third, it takes more power and more sophisticated antennas to send your signal to the satellite. Fourth, for the most part, you communicate using either SSB or CW. The two exceptions are the PACSATS, which require that you communicate via packet, and OSCAR 21, which has an FM repeater on board.

Communicating via a packet satellite is also similar to and different from communicating via packet. It's similar in that you connect to the packet station, but it's different because your packet station is always on the move, and you may not get all your traffic in one pass.

As with every sub-branch of the VHF+ frequency operations, satellite communications has its own language. Some of the terms and their meanings are given in the box "Amateur Satellite Terms" in this chapter. These are just a few of the words of "satellite speak" you'll use should you decide to communicate via satellites. You can refer to this list as we continue our discussion of satellite communications.

The Easiest Satellites to Work

Two Russian satellites, RS 10/11 and RS 12/13, are fairly easy to work through. Actually, there are four satellites—two per platform, hence the dual designation. However, only one satellite per platform is working at a time. These satellites operate in either Mode A, K, or T. In Mode A, the satellite is receiving a part of the 2 meter band and transmitting on a part of the 10 meter band. In Mode K, the satellite is receiving a part of the 15 meter band and transmitting on a part of the 10 meter band. In Mode T, the satellite is receiving a part of the 15 meter band and transmitting on a part of the 2 meter band. See Table 5-1 for the exact frequencies of operation for these satellites.

As you can see, for Mode K, you don't need a VHF+ station to work through the satellite. Because these satellites are low Earth orbiting, you can use your HF station and a dipole to talk with your fellow satellite operators. However, for the other two modes, you'll need to use a multi-mode 2 meter radio.

How do you work through a satellite? First, you need to find out when the satellite will be in the right position to hear you. This is where having a computer

Tom Webb, WA9AFM, operates this modest satellite station from the family room in his home. As Tom demonstrates, it does not take much to get "on the birds."

with the appropriate programs is almost a necessity. There are several programs available (check the *CQ Buyer's Guides* for more information). If you don't have a program, don't be dismayed. You can contact one of your local satellite enthusiasts and ask him to tell you when these satellites will be available. He will probably be glad to print out a few days' worth of predicted times of availability and expected modes.

Once you know when you can expect to hear the satellite and what mode to listen on, you can tune in for signals on the downlink frequency. You may just want to listen the first time around, so you can get a feel for what you hear.

The second time around, you'll probably want to be set up to transmit on the uplink frequency. If your license class permits operation within the uplink frequency, you can transmit either on SSB or CW. However, a note of caution is in order. If you are a United States licensee, *do not* transmit in voice on the 21.160 to 21.200 MHz uplink frequencies when RS 10 is in either Mode K or T. Voice privileges on these frequencies are *not* authorized to any class of U.S.

amateur radio operator. Also, if you have a straight Technician class license, you are not authorized privileges on this amateur radio band. Only Novices and Technician Plus and above are authorized to operate within this frequency sub-band.

Let's assume you know when the satellite will be in view. Because this book is about the VHF+ frequencies, and RS 10 is almost always available in Mode A, let's examine working through this satellite. You'll need a 2 meter radio—preferably a multi-mode model—and an HF radio capable of receiving on 10 meters, particularly between 29.300 and 29.500 MHz. Incidentally, obtaining a 10 meter radio should not be too difficult, because with the decline in sunspot activity, they are showing up for reasonable prices on the used market.

Because the satellite is a low Earth orbiting type, it doesn't take too much power to reach it, or much of an antenna. Therefore, you can probably get by with your outside-mounted vertical antenna. You can also use a vertical antenna for your HF radio. The omni-directional feature of this type of antenna will allow you somewhat continuous access to the satellite while it's passing overhead, without having to rotate the antenna.

Keep in mind, however, that should you get serious about working the satellites, you'll want to replace these verticals with multi-element cross-polarized antennas that can be rotated both in the azimuth and elevation planes. Nevertheless, for your first satellite experiences, you can use verticals successfully.

If you look at Table 5-1, you'll see that the uplink frequency for the RS 10 satellite is between 145.860 and 145.900 MHz. Typically, those who operate CW tend to stay close to the low end of the sub-band and those who operate SSB tend to stay toward the higher end of the sub-band. Notice that I do not mention FM. With its continuous carrier mode, FM is too draining on the satellite. Therefore, with the exception of OSCAR 21, no one uses FM. Consequently, if you only have an FM radio, you'll have to use it on CW by keying the mike, a bit awkward.

In order to hear the satellite's downlink frequency, tune your HF receiver to the beacon frequency of 29.357 MHz. As it gets close to AOS (Acquisition Of Signal) time, you'll want to tune the receiver up or down a bit because the signal may not be exactly on the designated frequency.

When listening for the beacon frequency, you may actually hear it some time before the satellite is scheduled to be overhead. This is because sporadic-E propagation may be carrying the beacon's signal to you from somewhere below the horizon. So don't let the seemingly early arrival fool you. Chances are that the information you have on its orbit is correct. Also, you won't hear your own signal being transmitted on 2 meters because it's highly unlikely that the same sporadic-E induced propagation existing on 10 meters is also present on 2 meters.

Once you hear the beacon signal during the overhead pass of the satellite, tune up the band and start listening for stations to work. You'll probably hear someone calling "CQ satellite" and signing his call. Note the frequency of that station. If, for example, the station is on 29.389 MHz, tune your 2 meter radio approximately 144.989 MHz. Now, sign your callsign and send a few dits while slightly rotating the 2 meter radio's VFO. Because of Doppler shift, you'll be changing frequencies—first higher as the satellite is moving closer to you and then lower as the satellite is moving away from you. Once you've found your signal, sign your callsign again because it's required, and because you'll know it's your signal you're hearing.

AMATEUR SATELLITE TERMS

AOS: Acquisition Of Signal. The point where you can start working through the satellite.

Apogee: The point in its orbit where the satellite is farthest from the Earth.

Ascending Node: The satellite is moving up from the horizon.

Azimuth: That point clockwise, on the compass, expressed in degrees, where you point your antenna in order to work through the satellite.

Descending Node: The satellite is moving down, or toward the horizon.

Doppler Shift: As it pertains to satellites, that change in frequency upward as a satellite moves toward you and downward as a satellite moves away from you.

Downlink: The signal from the satellite going to the ground, or Earth station.

Elevation: That point above the horizon, expressed in degrees (maximum 90°), toward which you point your antenna in order to work through the satellite.

Elliptical Orbit: A satellite traveling in this type of orbit is within a few hundred miles at perigee and over 18,000 miles at apogee. Because of this peculiar orbit, the satellite is in view for much longer, but because it is much farther away, it takes a more sophisticated station to work through it.

Footprint: That part of the surface of the Earth from which the satellite can receive signals and to which it can send signals at a given moment.

In View: This is a colloquial term used to indicate that you are underneath the footprint of the satellite through which you want to communicate.

Keplerian Elements: Named after Johann Kepler, the 17th cen-

tury German scientist who discovered properties of planetary motion. These elements appear to be a series of seemingly meaningless numbers, but are used in calculating the orbits of the satellites. They are available on many BBSs around the world after the launch of a satellite or Shuttle, or as in the case of the MIR space station, after an orbital correction.

LOS: Loss Of Signal. The point where the satellite has gone below the horizon and you are no longer able to work through it.

Low Earth Orbit: A satellite in this type of orbit travels close to Earth in the range of hundreds of miles. Because it is closer, it is easier to work through, but is only overhead for short periods.

Mean Anomaly: The relationship between where the satellite is and where it has been.

Mode: The uplink and downlink frequency pair, expressed in a single letter, on which a satellite is operating at a particular time.

Molynia Orbit: A high altitude elliptical orbit.

Mutual Window: Designation that indicates two stations at distant locations can work through the satellite at the same time.

Perigee: The point in its orbit when the satellite is closest to the Earth.

Period: The time it takes for a satellite to make one complete trip, or orbit, around the Earth.

Pointing Angle: The relationship, expressed in degrees, of the satellite's antennas to your QTH. Ideally, the pointing angle should be less than 30°.

Transponder: That part of the satellite which receives and then retransmits a portion of a particular band to another portion of another band.

Uplink: The signal from the ground, or Earth station, to the satellite.

RS SATELLITE FREQUENCIES

	RS-10	RS-11	RS-12	RS-13
Mode A				
Uplink	145.860-145.900	145.910-145.060	145.910-145.950	145.960-146.00
Downlink	29.360-29.400	29.410-29.450	29.410-29.280	29.460-29.500
Beacons	29.357/29.403	29.407/29.453	29.408/29.454	29.458/29.504
Mode A Robot				
Uplink	145.820	145.830	145.830	145.840
Downlink	29.357/29.403	29.407/29.453	29.454 —	29.504 —
Mode K				
Uplink	21.160-21.200	21.210-21.250	21.210-21.250	21.260-21.300
Downlink	29.360-29.400	29.410-29.450	29.410-29.450	29.460-29.500
Beacons	29.357/29.403	29.403/29.453	29.408/29.454	29.458/29.504
Mode K Robot				
Uplink	21.120 —	21.130 —	21.129 21.138	— —
Downlink	29.357/29.403	29.403/29.453	29.454 29.504	— —
Mode T				
Uplink	21.160-21.200	21.210-21.250	21.210-21.250	21.260-21.300
Downlink	145.860-145.900	145.910-145.950	145.910-145.950	145.960-146.000
Beacons	145.857/145.903	145.907/145.953	145.912/145.958	145.862/145.908
Mode T Robot				
Uplink	21.120 —	21.130 —	21.129 21.138	— —
Downlink	145.857/146.903	145.907/145.953	145.958 145.908	— —

Table 5-1. Frequencies of operation for RS satellites.

Note: Do not move the VFO on your HF radio. If you do, you'll get caught chasing the station you're trying to work up and down the band. As a result, you might "crash into" someone else's QSO.

Once you're relatively sure that you're close to the other station's frequency, try to make contact by signing both callsigns a couple of times. The repetition will give the other operator time to find you, once you've gotten his attention. Chances are, however, that you may have already gotten the other station's attention when you transmitted your dits and tuned your 2 meter radio to match his frequency.

If he hears you, the other operator will come back and give you a signal report (and/or maybe his grid locator), plus his QTH and name. You then respond with the same information.

You'll find the QSO is over fairly quickly, for two reasons. First, most operators, realizing that they have about 15 minutes at the most, are trying to work as many other operators as possible. Second, the other operator may be nearing the LOS (Loss Of Signal) time and trying to complete the contact.

There is a similarity to the uplink and downlink frequencies. The last two digits are always the same. This is true of all RS satellites. Knowing this makes it a bit easier to access any one of them.

ACTIVE AMATEUR RADIO SATELLITES

Name	Orbit	Type	Modes	Launch Year	Nationality
AO-10	Molynia	Analog	B	1983	Multi
UO-11	Leo	Educational	2m Receive	1984	UK
MIR	Leo	Manned (voice and digital)	2m FM	1986	USSR
RS-10/11	Leo	Analog	A (primary)	1987	USSR
AO-13	Molynia	Analog	B, S	1988	Multi
RS-12/13	Leo	Analog	K (primary)	1989	USSR
AO-16	Leo	Digital	J	1990	USA
DO-17	Leo	Educational	2m Receive	1990	Brazil
WO-18	Leo	Educational	J	1990	USA
LO-19	Leo	Digital	J	1990	Argentina
FO-20	Leo	Analog and Digital	J	1990	Japan
AO-21	Leo	Analog and Digital	B	1991	USSR and Germany
UO-22	Leo	Digital	J	1991	UK
KO-23	Leo	Digital	J	1992	Korea
Arsene	Geo Dr.	Analog and Digital	2m and S	1993	France
KO-25	Leo	Digital	J	1993	Korea
IO-26	Leo	Digital	J	1993	Italy
AO-27	Leo	Analog and Digital	J	1993	USA
POSAT	Leo	Digital	J	1993	Portugal

Table 5-2. *Active amateur radio satellites. (Information courtesy W5IU)*

Other Satellites

As I mentioned, there are nearly twenty satellites currently in orbit. Most of these are PACSAT satellites. Table 5-2 lists them as of the publication date of this book. (*Note: Those satellites listed as "educational" are "receive only," which means that you can only receive their signals.*)

In order to work through these satellites, you must have more sophisticated equipment and a special TNC capable of PSK and other modes. Your computer becomes even more important when working the PACSATs, as it will be controlling your transceiver's frequency in order to compensate for the Doppler shift and controlling your antenna rotators to track the satellite as it passes overhead. Operation through these satellites is beyond the scope of this book. If you are interested in pursing this aspect of the hobby, contact AMSAT for information on currently available publications. For a videotaped overview of working the satellites, you can check out *Getting Started in Amateur Satellites* (available from CQ Communications for $19.95, plus $3.50 shipping and handling).

The other voice satellites include OSCARs 10, 13, 20, and 21. All the OSCARs, except 21, use an inverting transponder. What this means is that your signal is reversed on the output from the input. Consequently, if you are on USB, your signal will come out on LSB. Additionally, if you are at the bottom end of the sub-band, you'll be listening for your signal on the top end of the sub-band. Table 5-3 lists the usual frequencies of operation.

When it's in the analog mode, OSCAR 21 is an FM repeater with an input (uplink) on 435.016 MHz and an output (downlink) on 145.9875 MHz. You'll use it in the same manner as a terrestrial crossband repeater—except there's no courtesy tone.

VOICE SATELLITE FREQUENCIES

Satellite	Mode	Uplink (MHz)	Downlink (MHz)
AO-10	B	435.030–435.155	145.825–145.975
	Beacon	—	145.810
AO-13	B	435.420–435.570	145.825–145.975
	J	144.425–144.475	435.990–435.940
	L	1269.351–1269.641	435.715–436.005
	S	435.601–435.637	2400.711–2400.747
	Beacons	—	145.812, 145.086
		—	435.651
		—	2400.325

Table 5-3. *Usual frequencies of operation for voice satellites.*

Again, as when using the PACSATs, with the exception of OSCAR 21, you'll fare better with more sophisticated equipment and a computer. However, with OSCAR 13, which is in an elliptical orbit, you'll have a great deal of fun operating through it because of the extended period of time (upwards of four hours!) that it's overhead.

As of the writing of this book, OSCAR 20 has been operating in the digital mode one week and in the voice mode the next week. Therefore, it's important, as mentioned earlier, to consult the various sources of information for current activity of each of these (both PACSATs and voice) satellites.

What About the Future?

Phase III D, which is scheduled to launch in 1996, will be an ambitious satellite that will be more powerful and in a higher elliptical orbit. Plans (as of this writing) include, among other things, a beacon in the 10 GHz band. Funds are being raised worldwide to support the development of the satellite and to pay for the launch. If you are interested in contributing to this challenging program, contact AMSAT at P.O. Box 27, Washington, D.C. 20044, or phone (301) 589-6062.

Looking into the next century, AMSAT planners are researching the possibility of a geostationary satellite. The plans for this satellite are included in Phase IV.

Incidentally, after the launch of Phase III D, the modes will take on a new designation. Gone will be the single-letter designation. It will be replaced by a two-letter designation, the first letter being the uplink mode and the second the downlink mode. For example, a satellite using 15 meters for an uplink and 2 meters for a downlink will be designated "KV" mode. The following are the letters and the bands for which they stand: "K" 15 meters, "A" 10 meters, "V" 2 meters, "U" 70 cm, "L" 23 cm, "S" 13 cm, "G" 5 cm, and "X" 3 cm.

Working Amateurs in Space

Both the United States and Russia have ham radio operators in space. The United States only does so during specified space shuttle launches, but the Russians always seem to have someone "up there" aboard the MIR space station.

The time of launch of a SAREX is announced well in advance. Publicity in *QST* and the *AMSAT Journal* provides plenty of notice. Bulletins issued by the ARRL and AMSAT will alert you to last-minute changes.

In order to work the astronauts, you'll need to know their Keplerian elements. Once known, you simply input them into your computer program. Again, if you don't have a computer, your local satellite guru will be happy to help you out.

You won't need much of a station to work the astronauts or cosmonauts. Your mobile station will do just fine. In fact, Troy Fehring, N5VIN, worked the shuttle from his tractor!

The frequencies of operation are published in the sources given above. It's important to remember two things. First, the shuttle uses 600 kHz offset, so don't transmit on the shuttle's transmit frequency, or you'll anger a lot of fellow hams. Second, astronauts are *very* busy. Just because the shuttle is overhead doesn't mean that one of the astronauts is available on the radio. Be patient!

For those of you interested in working the MIR space station, there is, as I said, a ham on board the MIR satellite almost all the time. You can either work the MIR station on voice or packet. Each cosmonaut signs a different callsign. The MIR station constantly makes orbit corrections. Therefore, it's necessary to check the above-mentioned sources of information to find out the latest information on the callsign and the updated Keplerian elements.

Recently, it has become somewhat easier to connect to the packet station. When the station is overhead, and once you pick it up, it's a matter of logging on as you would with a terrestrial packet station. It may take you a few tries, but keep at it and you'll get in!

As for voice communications, the cosmonaut who is comfortable speaking English may be more apt to be on the air. However, remember that the cosmonauts, like the astronauts, are very busy performing their tasks, and may not have much time to operate. Once again, be patient.

As this book was being prepared for printing, I received the following information. The latest Russian satellite, RS 15, which was due for launch in May 1994, may already be in orbit. The announced uplink and downlink frequencies are as follows:

Uplink—145.857–145.897 MHz
Downlink—29.357–29.397 MHz
Beacon frequencies—29.353 and 29.398 MHz

Satellite AO 27 now operates on FM with an uplink frequency of 145.85 MHz and a downlink frequency of 436.80 MHz. Satellite KO 25 now operates on AX.25 packet protocol using a standard TNC at 9600 baud.

Chapter 6

Packet Operation

In 1974 I took a course in FORTRAN, a programming language designed for business applications. I learned how to program in FORTRAN on an IBM 360 with a punch card reader as the input device. One of my assignments was to take a list of names and addresses and organize them into alphabetized address labels. Before the course was over, I had conned my instructor into letting me create my own QSL cards, completely filled out, ready for mailing, for extra credit.

It was this kind of imagination that fired hams who had their own Commodore, Timex, Heathkit, and Radio Shack computers in their shacks. They reasoned that if they could get their personal computers to send Morse code and RTTY, they could also get

them (operating at the breathtaking speed of 300 baud) to send traffic as well. Shortly thereafter, the Canadians, followed by U.S. hams, developed technology that required their respective governments to pass regulations authorizing transmission over airways. The Canadians launched a computerized radio communication system, which we know today as packet, in 1978. Then, in 1980, owing much to the efforts of the Tucson Amateur Packet Radio Corporation (TAPR), the FCC authorized the use of packet radio for U.S. amateur radio operators.

So what's it like to run packet radio? Running packet is a bit like using a modulator/demodulator (modem) to send files to a bulletin board (BBS) over the telephone line. However, instead of the telephone line,

Well-known contest operator John Guida, NJ1V, uses packet spotting to check for DX stations and multipliers during contests.

packet radio lets you send data over the air.

Packet radio gets its name from the fact that blocks, or "packets," of information are transmitted over the air one at a time. Error checking is performed, and when the receiving station is satisfied that all the information received is error free, it sends an acknowledgement to the sending station.

All this sending and acknowledging occurs without the intervention of a human operator. This automated setup alleviates the problems associated with errors made while copying the code, or propagation conditions that cause signals to drop out, thereby disrupting the flow of information on RTTY.

Who uses packet radio? Many different types of hams for many different reasons. Just about anybody involved in ham radio can find a use for packet. Rag chewers can spend hours in front of a computer or terminal and talk to their friends by punching away at the keyboard. Emergency traffic can be sent from the scene of an emergency. As an ARRL section manager, I have watched the bulk of messages, called radiograms, shift from voice and CW transmission to packet. DXers, who once used a dedicated FM repeater for local spotting of DX stations, now rely on spots on the DX packet cluster. Anyone on the DX end of a pileup can attest to the effect of being "spotted" on a packet cluster. All of a sudden, it seems as if hundreds of stations from one region are calling *you*. Contesters have acquired the packet cluster habit, so they can be alerted to necessary multipliers during contests. Finally, the computer enthusiast, who started it all in the first place, learns more and better ways to link the computer and amateur radio.

Packet radio has even opened one more door in the area of amateur radio for the handicapped. The deaf person, who previously found not being able to hear the code a hindrance, can now obtain a straight Technician class license and use packet to communicate. A lack of sight does not exclude the blind ham from operating packet. By adding a voice synthesizer card to the computer, a blind operator can "see" what's on the screen as it's read by the voice synthesizer. For example, Connecticut ARRL section manager Betsey Doane, K1EIC, who is blind, handles volumes of traffic via packet radio using her specially equipped computer.

Packet radio is like an answering machine for some hams. Through the use of packet bulletin board systems, or PBBSs, packet stations can send traffic to other stations even if the operator of the other station isn't home.

The Packet Station

What does a packet station consist of? As with a land-line-based system, your station will consist of a computer and a modem. However, because the signal is sent over the air instead of via the telephone line, you replace that line with a radio and antenna. And, because software (actually called firmware) is built into the modem, you can get by with using a terminal instead of a full-blown computer.

The modem used for packet differs from the modem used for landline communications principally in the built-in firmware. The packet radio modem is called a terminal node controller, or TNC. The firmware sets the protocol used in packet communications, so any packet station can effectively communicate with any other packet station. There's also a multi-node controller, or MNC, which is used by HF operators because it includes software that generates RTTY, CW, and other types of digital communications, such as AMTOR.

The protocol used, known as AX.25, was designed by the people at TAPR. It determines things such as abbreviations to save keystrokes when typing in command lines.

As you can imagine, given the various types of computers on the market, there's plenty of software available to make your favorite computer packet compatible, so let's set up your station.

First determine what you will use for a terminal. Will you use a terminal or a computer? If you use a computer, which will meet your needs? The IBM, or IBM-compatible, personal computer is the most popular. Consequently, you'll be able to find more software for this type.

Once you decide on a computer, you need to select a TNC. There are several TNC manufacturers, so check out several models and choose the one that suits you best. If you plan on using other digital modes, consider purchasing an MNC.

Now it's time to choose your radio. Because 2 meters is most popular for packet, it makes sense to use a 2 meter radio in your station. As this mode has grown in popularity, the inexpensive, used radios that were ideal for packet work have disappeared from the market. However, several manufacturers have stepped into the void, developing "data radios." Nevertheless, if you choose to operate on a band other than 2 meters, and plan on operating at a higher baud rate, make sure that the radio you choose is capable of performing higher speed (wider bandwidth) operations.

Your next consideration is the antenna. For the most part, an omni-directional vertical is the best choice because you find yourself connected to several different packet stations transmitting from several different directions all at once. However, should you want to talk to only one packet station, such as the DX packet cluster, a small vertically polarized Yagi antenna may be your best choice.

Once everything is hooked up, but before you get on the air, read your TNC operator's manual and a good packet book. I recommend *The Packet Radio Operator's Manual* by *CQ* magazine columnist Buck Rogers, K4ABT. You might even want to watch CQ Communications' video *Getting Started in Packet Radio*. Whatever resource you choose, be sure to use it with an eye to becoming familiar with the protocol involved in packet communications.

After you've read the books (and watched the videos) and everything is hooked up, you're ready to make contact with other packet stations. If you're on 2 meters, listen between 145.010 and 145.090 for the characteristic packet "braaackkk." Once you've properly tuned in one of the signals, you'll see information appear on your screen. From here it's just a matter of connecting and jumping in.

The Southeast Packet Cluster™ A Great Way To Shop For VHF+ DX

I've explained some of the benefits of HF DX spotting. However, in recent years VHF+ DX spotting has come into its own. The Southeast Packet Cluster is an example of one successful VHF+ spotting network.

For several years the Southeast Packet Cluster system has provided opportunities for DX spotting for operators in Alabama, Georgia, North Carolina, South Carolina, Tennessee, and parts of Kentucky. There are around 400 users in 23 nodes who have worked many a DX station thanks to the spotting provided by the packet cluster.

As interest in VHF+ DXing increased, so did the spotting of VHF+ DX on the network. As spotting increased, the system overloaded. The increased workload bogged down the system, so at times it took upwards of 20 minutes to span the cluster. Clearly, a solution was needed.

A couple of years ago Bart Fay, K4CEF, discovered that a (then) new distribution list (or DISTRO) feature could be used to send mail or announcements to cluster users who were interested in a particular topic, without having to send the information to everyone

using the commands "S ALL" or "ANN/FULL." Bart mentioned this find to Bob Striegl, KA2DRH, an avid VHF+ operator.

At once the two saw the potential for providing VHF+ DX announcements for those interested, without imposing these spots on the entire system. Bob contacted stations that frequently showed VHF+ spots and asked if they were interested. A list was developed which ultimately grew to 75 members. At the same time Bart set up a DISTRO list named "VHF" on their node. Now, if someone wants to identify DX on the VHF+ frequencies, he types ANN/VHF (message) and the message only goes to the other VHF+ users.

As the system was refined, the requirement that each node have the entire list was eliminated. Thanks to software advances, this was accomplished by having stations that define a particular node as their home node included on the VHF DISTRO list indicating that node. Bart now edits the VHF DISTRO list remotely at each node when a new station wants to be added or an old station drops off.

Bob maintains a backup file and hard copies of the list and occasionally solicits new members, and dispenses user directions. As new nodes join the cluster, they are given information on the list and the opportunity to form and edit their own. They are advised to keep Bob and Bart informed, so the master list can be maintained.

One further refinement has been made with regard to mail distribution. Bart and Bob found that with more than 75 members, doing an "S VHF" command generated too many mail messages. This, in turn, clogged mail distribution throughout the cluster. To solve this problem, Bart created a dummy user at his node and named it "VHFMAIL." Now, all mail for the group is sent to the dummy user first. A distribution list with embedded mail routing is appended to the message, and one copy is routed to each node. There, the message is recopied as many times as necessary and routed to all the node users.

Bob comments, "Bart's intimate knowledge of the topography of the cluster and inventive mind made the VHFMAIL a viable way of communicating information on grid expeditions and equipment buy/sell messages, and a great way for me to reach everyone on the list when I need to." Bob adds that he is constantly updating the group list and passing along that information to Bart for incorporation.

Bob and others have found the list very helpful for adding to their grid, state, and country totals. Bob also reports that he has made many contest skeds and even a few meteor skeds from the list. He adds that, to date, he has worked nearly everyone on the list. If you're interested in getting on the list (or working Bob) contact him at Rt. 10 Box 161A, Athens, Alabama 35611, or call him at (205) 729-1429.

Chapter 7

Amateur Television Operation

When I first spent time around my mentor, Bert Adams, K6BTO's shack, I noticed that he had a television in the corner. I jokingly said something about Bert being so bored with ham radio that he kept a TV around to break the monotony. Bert pulled me up short by telling me that yes, indeed, he watched the TV, but he watched his fellow hams! The year was 1960, and this was my first exposure to amateur television, also known as ATV.

Back in the 1960s the ATV operator was also the weak signal operator. Today's ATV enthusiast, however, is anyone with an interest in television communications. In recent years ATV operation has been given a significant boost by the relatively easy access to camcorders. In fact, ATV operation is one of the fastest growing specialties in the VHF+ world.

Amateur television has grown beyond the boundaries of sending pictures of one's shack to other stations. ATV transmitters are now mounted in balloons, model airplanes, and robots. ATV is also used to retransmit weather maps,* and crowd and disaster scenes for public officials. In fact, ATV has become an integral part of the Tournament of Roses parade in Pasadena, California, and is used to handle crises along the route. Focusing ATV in on a particular problem, such as a disabled float, lets emergency service personnel access the situation and dispatch the appropriate equipment and/or personnel to fix it.

Interest in ATV, both slow and fast scan, has grown steadily over the years. Fast scan ATV is like that found in commercial broadcasting. Because it takes an enormous amount of spectrum space to transmit a fast scan ATV signal, it is only permitted on the 70 cm and above ham bands.

As in the 1960s, the principal band for ATV operation today is 70 cm. Most ATV operation can be found on 434.000 or 439.250 MHz. ATV can also be found on the 33 and 23 cm bands.

What's different today is the use of ATV repeaters —particularly crossband repeaters. Using an input frequency of either 434.000 MHz (in high population areas) or 439.250 MHz (in the midwest or low population areas), repeaters retransmit ATV signals within the 70 cm band or crossband on either the 33 or 23 cm ham bands. (See the Appendix for exact frequency allocations for these bands.)

While the vast majority are within-band repeaters, more of the new repeaters are crossband. There are several advantages in using crossband repeaters. First, desensing at the repeater site is nearly eliminated. Second, the operator transmitting a signal can see the signal and doesn't have to rely on others to make adjustments to it. Third, use of a crossband repeater frees up the other 70 cm ATV frequencies for simplex operation.

Most ATV contacts are not initiated by putting a

*Retransmission of near real-time weather RADAR maps is in a "gray area" of FCC jurisdiction because it is technically rebroadcasting another service, even though it has gone through computer processing beforehand. Therefore, about the only time a weather map is retransmitted is under the auspices of the Radio Amateur Civil Emergency Service (RACES), and then only during severe weather for the purpose of coordinating spotters and relaying vital information back to the National Weather Service and appropriate government agencies. There is one exception in the FCC rules that many ATV repeater operators take advantage of. This is the retransmission of space shuttle video and audio from NASA Television. Incidentally, this signal can be found on the Spacenet 2 satellite, transponder 5, C Band, 69 degrees west, on 3880 MHz, horizontal polarization, with the audio on 6.8 MHz.

35

Deitz Bigelow, WB5ADN, an avid ATV operator, is also a weak signal VHF+ operator. Deitz belongs to the ATV Experimenters Society, a group of nearly twenty ATVers in the Oklahoma City area. You can find ATV operation in almost all major metropolitan areas.

signal on the air and calling "CQ ATV." Contacts are arranged when someone gets on the ATV coordinating frequency and announces that he is going to put a video carrier on the air. Incidentally, the calling frequency depends on which ATV frequency you are using. Generally, if you use 439.250 MHz, the calling frequency you'll use is 144.340 MHz; if you use 434.000 MHz, the coordinating frequency is 146.430 MHz. The principal reason for the difference in these two frequencies is to prevent the third harmonic of the 2 meter signal from interfering with the ATV signal.

Once an ATV signal is on the air, the operator of the transmitting signal will switch from the coordinating frequency to the voice, or aural, subcarrier of the TV signal for his audio transmissions. Meanwhile, the calling frequency continues to be used, only now by others for communications with the operator transmitting the signal.

As with any other type of QSO, a form of signal report exists for ATV contacts. ATVers use a rating system of 0 to 5, preceded by the letter "P." For example, a signal that just shows the sync bars in the snow is rated "P0"; a very snowy, almost imperceptible signal is a "P1"; and a clear signal with no snow is a "P5." Some operators go so far as to split the sig-

nal reports into tenths. Thus, a signal with just a small amount of snow is a "P4.5." Incidentally, a P5 report represents about 150 to 200 microvolts of signal in the receiver.

Equipment Requirements for ATV

What does it take to get on the air with ATV? Surprisingly little. There are 10 watt transceivers available for around $500. These transceivers can be obtained with a crystal for one or two of the popular ATV frequencies and are set up to convert the receive signal to television Channel 3.

This use of your television assumes that you'll be watching the ATV signal being transmitted on the 70 cm band. However, if you're using a crossband repeater, you'll have to obtain a converter for the band of the repeater output frequency. These converters connect directly to the TV from the antenna. You can use a video switch to change from the converter output to the transceiver output. If you don't wish to purchase a converter, you can put a second TV to use as one!

Your biggest expense will be the camera. If you can live with monochrome, surplus cameras are available from a variety of sources. However, if you want color, and don't mind tying up your camcorder while you're on the air, you can use that as a dual-purpose unit.

You have quite a few antenna choices, and you'll need to know a bit about your area before you purchase one. If you're going to work simplex and plan to use 439.250 MHz, then you'll want a horizontally polarized antenna to avoid the increasing problem of encroachment from outputs of vertically polarized FM repeaters. If you're going to use 434.000 MHz, you'll want a vertically polarized antenna to avoid interference from weak signal stations that use 432.100 MHz. However, if you're using a repeater, you'll want to know its polarization. Most repeaters are vertically polarized and use commercial high-gain omni-directional antennas. Check with nearby ATV enthusiasts to learn your local repeater's polarization.

If you're trying to work DX, you'll want a horizontally polarized antenna, because most ATV DXers come from the weak signal community and use horizontally polarized antennas. However, if you want the flexibility of both vertical and horizontal polarization, a satellite antenna that enables you to switch between the polarizations will serve you well.

For the 2 meter transceiver you use for the coordinating frequency, you'll want a vertical antenna with some gain. This will put you in a round-table with other ATV operators, and an omni-directional antenna will preclude the need to rotate your antenna to hear each of the other stations.

What if you're not quite ready to make the commitment to ATV? If you don't want to invest in a transceiver or down converter, but simply want to be an SWL, you can use your cable-ready TV as a receiver. To do so, you'll need to put your television in the "cable" mode. Many televisions have a switch that changes between cable channels and channels received by the antenna input. I have two cable-ready televisions and each is programmed differently. The Sony has a switch that says "Cable, Off, On." The Emerson uses remote control programming to select between "TV" and "Cable TV." Consult your owner's manual to find out how to program your set for cable reception.

Once you've set up your TV, you can tune directly to Channel 60 if you're using 439.250 MHz as your ATV frequency, as the frequency of that channel is also 439.250 MHz. If you're using 434.000 MHz as your ATV frequency, tune to cable Channel 59. However, if you have a "Fine Tuning" adjustment, you may have to make a slight adjustment to your television, because the cable frequency is 433.75 MHz (otherwise, don't worry about it, because your television may have an automatic fine tuning, or AFT, circuit that will search for the best reception of the signal the television picks up). If you're in an area with a 70 cm repeater that uses 421.25 MHz for an output, you can tune to Channel 57, which is on that frequency.

Now, it's simply a matter of finding out what frequency is used by ATV enthusiasts in your area, finding out where the signal is coming from, connecting the right kind of antenna, and rotating it. There's nothing to stop you from joining in on the round-table discussion on the coordinating frequency. Watch out, though, as you just might get hooked!

What's in the future for amateur television? ATV operators are discovering ways to send the digitized signal from computers over ATV. Infrared cameras that began to appear on the surplus market following Operation Desert Storm are giving ATV operators the ability to make night-scene transmissions.

Ballooning has come a long way from the days when my high school buddies—Scott Morton and Richard Thorn (both of whom held Novice class licenses for a while)—launched a balloon with a CB walkie talkie wired to transmit a chirpy tone, so they and I could track it using their CB radios and my general-coverage receiver. Now, balloons that go up to over 100,000 feet and show the Earth and edge of space over a 500 mile radius are launched with GPS receivers that translate and superimpose the coordinates with the video signal, so they appear across the bottom of the screen. So much for hidden transmitter hunting!

Uses of amateur television are only limited by your imagination. Perhaps someday, like my mentor Bert Adams, K6BTO, I'll be watching you via television.

Chapter 8

Weak Signal Operation

The term "weak signal" is something of a misnomer, because it implies that all this VHF+ specialist does is spend time looking for weak signals. While that is somewhat the case, often the "weak signals" heard are quite loud. Nevertheless, the weak signal specialist enjoys the challenge of copying signals much weaker and oftentimes thought impossible to hear on the VHF+ frequencies. Let's look at each of the VHF+ frequency bands and see what drives the weak signal operator.

6 Meters

The 6 meter band is sometimes thought of as the last of the HF bands, just as it is sometimes considered the

second VHF band because it so often shares the characteristics of both HF and VHF propagation and because 10 meters is sometimes considered the first VHF band. Throughout the year, but in particular during extended periods that occur twice a year, the VHF+ operator encounters sporadic-E propagation. This type of propagation allows for the completion of contacts in the 1,200 to 1,300 mile range (and sometimes much more), depending on whether or not it's single- or multiple-hop sporadic-E. Occasionally, multiple-hop sporadic-E propagation can make possible contacts to distant (5,000 miles) parts of the world. Signals from single-hop sporadic-E propagation can be quite loud.

Tropospheric (tropo) scatter propagation occurs

This beautiful 6 meter EME array was engineered, designed, and built by Bob Magnani, K6QXY, and Al Ferrera, WA6MXA. (Photo courtesy K6QXY)

At the end of the rainbow is this golden 32 foot dish which is used by Marc Thorson, WBØTEM, in Akron, Iowa for EME on 70 through 13 cm. (Photo courtesy WQØP)

almost every day of the year. Meteor scatter is also usually present, although it's better during the months of July through December. The best time to catch meteor scatter is during the early morning hours. Both tropo and meteor scatter propagation permit contacts of 1,000 miles, or more. Signals from tropospheric scatter are often weak, but signals from meteor scatter propagation can burst through very loudly for periods of a few seconds up to a couple of minutes.

Auroral propagation is occasionally present, mostly in the northern latitudes. Hams in these areas have been known to make contacts using auroral propagation in excess of 2,500 miles by propagating their signals along the auroral oval. Often, signals via auroral propagation are weak and watery sounding.

During peaks in the sunspot cycle, F_2 propagation is present. The height and length of the sunspot cycle peak determines how often this form of propagation occurs. When it is available, this mode of propagation permits contacts of 6,000+ miles or more. Transequatorial (TE) propagation is related to F_2 propagation. This form of propagation permits contacts between stations that are equidistant from each other across the equator. Signals sent and received via F_2 and TE propagation are usually weak, but are sometimes quite loud and intense. I once watched my "S" meter for half an hour while a station from Australia kept the needle at the 20+ dB mark.

VHFers have been bouncing signals off the moon via the 6 meter band for some time. While interest in EME waned somewhat during the peak of Solar Cycle 22, it's enjoying a resurgence following the cycle's decline. During sunspot decline weak signal operators who are hooked on worldwide communications have found that EME communication is about the only way left to make these kinds of contacts. Signals from the moon are usually quite weak, and you need high power and big antenna arrays to communicate regularly via this mode. For the most part, this is the form of propagation that earns the weak signal operator his reputation.

2 Meters

Although originally designated an experimenter's band (see Chapter 4), 2 meters has become the workhorse for the repeater operator. Nevertheless, the weak signal operator makes very good use of 2 meters for many weak signal contacts.

During times when sporadic-E activity peaks on 6 meters, 2 meters may occasionally host such propagation. However, it is short lived and only rarely will permit contacts in excess of 1,300 miles. As on 6 meters, signals can be quite intense. Sometimes, when sporadic-E is dying out on 6 meters, propagation via field aligned irregularities, or FAI, pops up. Signals via this propagation mode are weak and watery.

Tropospheric propagation is a regular event on the 2 meter band. As a repeater operator, you occasionally may be exposed to this form of propagation as stations from distances of 150 or more miles come through your repeater. These signals can be fairly loud on the repeater, and they also can be consistently loud to the weak signal operator. Often, however, they are weak, but steady.

Meteor scatter propagation is quite popular on 2 meters. Contacts around 1,000 miles or so can be made regularly. Meteor scatter enthusiasts know when specific meteor showers occur and plan their activities accordingly (see Chapter 11). Signals from meteor scatter are often weak, but occasional pings can last from a fraction of a second to as long as a minute, or so.

Auroral propagation is present here, but not like it is on 6 meters. Contacts of 500 to 800 miles are possible. The signals usually are weak and buzzy sounding.

While F_2 has never been reported on 2 meters, TE propagation has been documented on very rare occasions. Signals are extremely weak, and sophisticated equipment is necessary to work this propagation mode.

EME is quite popular on this band. As on 6 meters, signals are quite weak.

135 cm

Following the loss of the bottom 2 MHz of this band a few years ago, weak signal activity, which had a designated portion of that lost section, almost ceased to exist. Because of so much uncertainty, weak signal operators lost interest and quit the band. However, on February 1, 1994, the FCC designated the bottom 150 kHz as experimental, paving the way for a resurgence of activity by the weak signal operators.

About once every ten years, or so, someone reports contacts via sporadic-E. It is very rare, indeed. When it occurs, signals are weak to fairly strong, but they don't last long.

Tropospheric propagation occurs quite regularly on 135 cm. Signals are moderately strong. Meteor scatter contacts are made on this band, but with more difficulty than on 2 meters. Signals are weak, and propagation lasts for fractions of a second to just a few seconds. Auroral contacts are possible, but much less often than on 2 meters. As on 2 meters, the signals are weak and buzzy sounding. Transequatorial contacts have been documented, but only on very rare occasions.

Contacts via the Moon can be made anytime. However, because of the previous uncertainty of the future of weak signal operations on this band, EME on 135 cm had mostly been discontinued. Nevertheless, with the authorization of the new sub-band, EME, as well as all weak signal operation, is enjoying a resurgence.

70 cm

Seventy cm is a workhorse band for the weak signal operator—principally because of tropospheric and EME contacts. Tropospheric propagation occurs regularly. Signals are moderately strong. Meteor scatter contacts are made, but with much more difficulty than on 135 cm. Auroral contacts have been made, but rarely. Transequatorial contacts have been documented, but are also very rare. EME contacts are made quite often; they are still very weak, but somewhat louder than on the lower bands.

33 cm

The 33 cm band is another band with a clouded future. As of the writing of this book, several services are vying for frequency allocations within this band. Because of this uncertainty, only weak signal operators and amateur television enthusiasts make regular use of it. However, even that use isn't as frequent as it could be.

Tropospheric propagation occurs regularly. EME contacts have been made and are quite possible, but aren't made often, due to the lack of activity.

23 cm

The 23 cm band enjoys a lot of popularity with weak signal operators. Tropospheric contacts are made regularly when the band is open, and EME contacts are made all the time.

13 cm

As commercial equipment becomes increasingly available, 13 cm will gain in popularity. As it is now, only weak signal operators who build their own equipment, or put together their own kits, are on this band. Tropospheric operation is regular when the band is open. EME propagation is quite possible, although only a handful of operators pursue it regularly.

9 cm

Again, because of the lack of equipment, only a few weak signal operators use the 9 cm band. Tropospheric propagation is the principal mode of communication. EME contacts can be made, but few operators have the equipment to do so.

5 cm

The 5 cm band also suffers from lack of activity due to lack of equipment. Tropospheric propagation again is the principal mode of communication. EME contacts can be made, but only a very few operators have the equipment.

3 cm

Owing to the availability of surplus radar transmitters, detectors, and Gunnplexers, 3 cm is quite popular for short-range contacts. Occasional "over-the-horizon" contacts are possible. EME contacts have been made, but by less than a dozen people.

1.2 cm and Above

The 1.2 cm band is increasing in popularity with those operators who first started out on 3 cm. Again, homebrew equipment is being used for short-range contacts. It is, however, the highest band where any regular experimental communications are taking place. Propagation is affected by factors such as water vapor and oxygen.

Experimental work is being done on 47 and 145 GHz by operators in Europe. A couple of operators in the U.S. (Tom Williams, WA1MBA, and Jim Mead, WB2BYW) are also experimenting on 145 GHz. Above this band, except for laser, no regular or experimental work is taking place.

Laser

Laser experimentation occasionally takes place, but not on any regular basis. It appears that the only challenge occurs when someone has broken someone else's record. Operators then drag out their equipment and attempt a new record. The only regular laser communications take place during contests, and then principally for increasing a score with multipliers from this operation.

Part II
Weak Signal Communications

Chapter 9

Sporadic-E and FAI Communications

From the middle of May through the end of July, and again in late November to early January, sporadic-E propagation appears more frequently on the VHF+ frequencies in the Northern Hemisphere. This type of propagation occurs when there's a sporadic ionization of the E layer (the layer between 60 and 70 miles, or 100 and 120 km, above the Earth). The ionization takes the form of clouds of ionized gases that move, grow larger and more intense, then shrink and dissipate. These ionized clouds appear in the late morning and late afternoon local time. Late afternoon ionization can last until well after sundown and sometimes overnight.

For example, one evening a couple of summers ago, I was listening to a local net on a 2 meter repeater. A new ham checked in at nearly 2300 local time to exclaim that 10 meters was "just hopping with signals." I picked up the clue and immediately turned on my 6 meter radio. The last of the signals on that band faded at around 0100 local time. I then switched to 10 meters, where propagation continued for another hour, or so.

Sporadic-E ionization propagation has properties similar to other forms of E-layer propagation. Depending on the density of the ionization, there's a critical frequency (CF), a lowest optimum frequency (LOF), and a maximum usable frequency (MUF); over this range of frequencies usable signals are refracted back to Earth. The CF refracts straight back down those signals sent straight up. The LOF is the lowest frequency that will sustain propagation. The MUF is the maximum frequency that can sustain propagation. The MUF is usually about five times the CF. Though the LOF of a sporadic-E event has been detected as low as the 20 meter ham band, interest in the low end hasn't been as strong as interest in the

MUF of a particular opening. This is partly because it's very difficult to distinguish sporadic-E propagation from other forms of propagation happening at the same time.

As VHF+ operators, we're interested in knowing how high in frequency we can use this mode of propagation. Obviously, 6 meters is the VHF+ band that benefits most from sporadic-E. However, 2 meters, and on very rare occasions 135 cm, has experienced sporadic-E propagation. The historic 220 MHz contact between Bill Duval, K5UGM (in Irving, Texas), and John Moore, W5HUQ/4 (near Jacksonville, Florida), on June 14, 1987 at 1544 UTC is the only known documented sporadic-E contact that has ever been made on that band.

Sometimes clouds of ionization may be in just the right places to cause double-hop sporadic-E propagation. Clouds are less often in enough of the right places to induce triple-hop, or greater, sporadic-E, but this does occur often enough. For example, almost every summer, hams living on the upper eastern coast of the United States experience a few days of multiple-hop sporadic-E into Europe. Sometimes this propagation extends as far south as the Carolinas. Occasionally, there are openings between the west coast and the far east. Once in awhile, stations in parts of the United States are able to complete their WAS awards when either Hawaiian or Alaskan stations come in via a multiple-hop sporadic-E path.

For the 6 meter operator, the most popular propagation mode is sporadic-E. It affords regulars to the band opportunities to talk across the country and, on occasion, into foreign countries. It's the most ready avenue (and probably the first introduction to) DX for the VHF+ operator. Because of its nature, it doesn't take much power to work stations.

In the 1992 CQ WW VHF WPX Contest sporadic-E gave a number of stateside stations an unexpected contact with YW5N. Shown here are the operators of YW5N: (back row) YV5LIX, YV5GE, YV5LDL, and YV4BHJ; (front row) YV5NN, YV5NFX, and YV5JCB. (Photo courtesy YW5N)

What causes sporadic-E? No one seems to know, exactly. Wind shear gets most of the blame because it seems to cause a bunching up of ions, creating a cloud of ionization. However, the wind-shear theory looks questionable when sporadic-E is present without it. The "old wives' tales" of amateur radio have also associated sporadic-E propagation with thunderstorms, aurora, and meteor showers. While there is some connection between E-layer propagation, aurora (more commonly known as auroral-E), and meteor showers, the jury is still out when it comes to thunderstorms.

Will sporadic-E occur in your neck of the woods? It depends on where you live. Sporadic-E seems to be most prevalent in the southeastern part of the United States.

How can you learn to depend on sporadic-E propagation? Start by listening to 10 meters. If you hear exceptionally loud signals from an area that's not too far from you geographically, you may want to turn on your 6 meter radio—especially if the "skip" starts to shorten considerably. Once on 6 meters, you can begin listening to see when the skip shortens on the band. You might even find some propagation on 2 meters. Of course, such propagation doesn't always occur. In fact, it only occurs about 10 percent (or less) of the time that it occurs on 6 meters.

Doug Allen, W2CRS, has his own method of detecting sporadic-E on 2 meters. An active commercial FM DXer, Doug says he keeps an ear on a clear fre-

quency near the highest frequency on the commercial FM spectrum (108 MHz). When he hears signals coming in from a distance, he immediately turns on his 2 meter radio and starts transmitting. Oftentimes, he "creates" his own opening. Doug says he can detect all sorts of propagation, from sporadic-E, to tropo, to meteor showers, by listening to a clear frequency on the 2 meter band.

During high sunspot activity, when F_2 type propagation is present more often, sporadic-E has been known to contribute to the lengthening of a path of a propagated signal. For example, twice in January of 1993, DX contacts appeared to have been assisted by sporadic-E. Stations in New Zealand were working stations in Arizona when a path opened to Oklahoma for the ZLs. At the same time, a sporadic-E path existed between Arizona and Oklahoma. It looked liked the signal took a ride on F_2 to Arizona, then hopped a ride the rest of the way via sporadic-E. Within a few days of that opening Dave Batcho, N5JHV, experienced an unusual opening to central Europe. On both ends (stateside and European) of the circuit, sporadic-E propagation was reported (in the U.S. between the northeastern and the southwestern parts of the country and in Europe between England and central Europe).

Sporadic-E propagation presents an opportunity to communicate with distant amateur radio operators, while running marginally equipped stations using very low power. Sometimes you'll find that the signal strength of the station you're working is so intense, you'd think it was local. But you need to work that station quickly, because sometimes an opening closes as soon as it appears. Such is the nature of the clouds that make up sporadic-E propagation.

FAI Propagation

Field aligned irregularities, or FAI, propagation is found on both 6 and 2 meters of the VHF+ bands. Experiments have shown that FAI can exist on 135 cm and maybe even on 70 cm. While this form of propagation has been around for some time (experiments conducted by Stanford University in the 1950s confirmed its existence), it wasn't until Tom Kneisel, K4GFG, documented his 1978 2 meter contact with Dave Ternet, KP4EOR, that amateurs more clearly understood FAI propagation.

In an article Tom published in the January 1982 issue of *QST* entitled "Ionospheric Scatter By Field-Aligned Irregularities at 144 MHz,"[1] Tom tells how he set up a sked with Dave over successive nights during the summer of 1978. When finally alerted to the fact that Dave was on the band (although off Tom's frequency by 3 kHz) by a phone call from Doug Welcker, WB4KGY, Tom and Dave both discovered that the peak of their signal was 20 degrees north of the direct (or great circle) path. Tom conducted further schedules with Dave and also with stations in Texas. Each time he observed that his beam heading was north of the great circle path.

Tom also researched the work conducted by the Stanford University scientists. Through his experiments and research Tom discovered that, approximately half of the time, FAI occurred right after the peak of intense sporadic-E activity along the same path. He found that the FAI often occurred as the sporadic-E was starting to die out and would last as long as a half hour or so after the sporadic-E propagation was gone. Tom further observed that during the months of the year when little or no sporadic-E was present, he was unable to make FAI contacts. Finally, Tom noted that he could only make FAI contacts at night following nighttime sporadic-E events.

In the article Tom described the signal quality as fluttery, like aurora, but not as disruptive or buzzy. He says that there are frequency components of 30 to 100 Hz present, but they don't destroy the signal like those found in aurora. Because of this, SSB contacts are possible. He reported that at times parts of minute-long transmissions were clear and free of the clutter, an observation that he couldn't explain.

Tom reported that stations with limited equipment can occasionally work FAI, but high power and/or a good antenna make all the difference.

If you're interested in this form of propagation, wait until you hear the dying embers of a very intense sporadic-E opening on 6 meters, turn on your 2 meter radio, point your beam between 15 and 25 degrees north of the direct path you were hearing on 6 meters, and make some noise. You just might make some contacts using this exotic mode of propagation.

References

1. Thomas F. Kneisel, K4GFG, "Ionospheric Scatter By Field-Aligned Irregularities at 144 MHz," *QST*, January 1982, pp. 30–32.

Chapter 10

Communicating Via Tropospheric Propagation

It's early morning and you step out onto the front porch. You're immediately blasted with oppressive humidity and the hot summer air. You know that it's going to be another scorcher.

You turn on your handheld and listen to the various repeater frequencies in your area. You accidentally tune to a non-local frequency and are surprised to hear signals. You listen closely and determine that the repeater signal is coming from a distance, possibly as far as 100 miles. The signals you're hearing are being propagated via a tropospheric enhancement.

Tropospheric enhancement can occur any time there's a temperature inversion. Generally, summer is the best time. Temperature inversions can occur any time one air system is overrun by another. As a result of this overrun, signals are trapped relatively close to the ground and are propagated over greater than normal distances—occasionally as far as 1,000 miles.

Tropospheric propagation can also happen over water. The distance covered may be in excess of 2,000 miles. When tropo occurs over water, it travels through a duct, like a tube. This tube has been described as a runaway hose moving in slow motion. Operators who have experienced this ducted communication report increasing and decreasing signal strength as the "end" of the tube moves from their location to another and back again.

The MUF and LOF of this phenomenon behave quite differently from HF propagation. When tropospheric propagation makes its initial appearance, it may first affect the 70 cm ham band. Sometimes the signals are heard first on the 23 cm ham band, with the MUF of the propagation increasing to as high as the 5 cm ham band. However, an event like this is extremely rare.

Hams have been aware of tropospheric enhancement for a very long time. Experiments on 60 MHz in 1934 by Dr. C. F. Brooks, who used the experimental station W1XW at the Blue Hill Observatory near Boston, and Ross Hull, located at West Hartford, Connecticut, made amateur radio operators aware of this form of propagation.[1] However, it wasn't until 1957 that John Chambers, W6NLZ, and Tommy Thomas, KH6UK, conducted experiments based on airline pilots' reports of being able to hear the Honolulu tower once in the air after takeoff from a west coast airport.

These two tropo pioneers made their first contact on 2 meters on July 8, 1957 and set the next record on June 22, 1959—this time on 125 cm. Paul Leib, KH6HME, in Hawaii, and Jack Henry, N6XQ, operating from Baja, California, later stretched out the distances for contacts on these bands.

It wasn't until July 18, 1979 that a contact was made on 70 cm. Louis Anciaux, then WB6NMT, now KG6UH and HL9UH, who then lived on a ridge of Point Loma in San Diego, had an excellent view of the Pacific Ocean. When he heard the 70 cm beacon located on the side of the Mauna Loa volcano, he called Paul Leib, KH6HME, on the phone. He then had to wait five hours for Paul to get off from work and up to the site in order to complete the first contact.

Five years later the first contacts were made on yet another VHF+ band. This time, on June 24, 1984, Chip Angle, N6CA, heard the 23 cm beacon on Mauna Loa that he had built and sent to Paul a few years earlier. Chip alerted Paul and made plans to drive to Palos Verdes. Meanwhile, Paul made his way up the side of Mauna Loa. The two made contact with relative ease.

In 1991, Paul and Chip were at it again. On July 28, Chip, again operating from Palos Verdes, made contacts on the 9 and 5 cm ham bands. Most recently, August 23, 1993, these two once again completed a record-setting contact over the Pacific Ocean—this time on the 33 cm ham band.

Records were also being set between Australia and New Zealand, using the same phenomenon over water. However, no contacts have been completed to date on 13 cm, or on frequencies above 5 cm, although attempts are being made between California and Hawaii, and between Australia and New Zealand.

Over-land records have also been set. As recently as September 16, 1993, Al Ward, WB5LUA, and Tom Whitted, WA8WZG, completed a 952 mile contact on 2304 MHz, setting an over-land distance record for that band.

During the 1992 September VHF QSO Party Sam Whitley, K5SW, worked the W2SZ contest group on 222 MHz, stretching the over-land distance for that band out past the 1200 mile mark.

How to Tell When to Operate

Besides stepping outside and noticing that the weather is "stuffy," how else can you determine that tropospheric propagation conditions exist? First, listen on the bands for distant beacons (see the list in the Appendix). If you hear these beacons, the band is open. You can also determine that conditions exist by checking the weather map. If you're sitting in the middle of a high pressure area that's being closed in upon by a low pressure area, chances are good that you'll be experiencing tropospheric enhancement.

When tracked over a 24-hour period, tropospheric enhancement seems best during the early morning hours just after sunrise. During these early morning hours conditions occur that refract VHF+ frequency signals along the surface of the Earth. Tropospheric conditions can return after sundown when the air is being cooled by the absence of the sun.

Another way to determine if there are tropo openings is to listen for a DX station on your local repeater. As ARRL Oklahoma Section Manager, I've traveled to many hamfests and club meetings throughout the state. On one such trip to Lawton, I was listening to the local 146.82 repeater around 9 a.m., when a couple of Texas operators checked in and reported that our operators were breaking into their repeater. Being the VHF+ weak signal operator I am, I immediately tuned my radio to 144.200 MHz and started calling CQ. I was first called by Howard Hallman, WD5DJT, who couldn't quite get a fix on me and kept asking me to repeat my grid locator. Thinking that I could help him copy me, I called CQ again, this time repeating my grid locator several times. I was answered by Pat Rose, W5OZI, in EM00, nearly 400 miles away! We both were quite startled to make the contact, because I was running only 25 watts into a SQLOOP antenna. By the time I completed the contact with Pat, Howard had gotten a fix on me and called again. We completed our contact just before tropo conditions deteriorated.

Over the years of my writing my column in *CQ* magazine, I have received many reports of tropo-induced contacts. Sometimes signals are very weak and sometimes they are very intense. On some of these occasions operators have experimented with reducing power to the absolute minimum their stations will run.

At times these conditions will last for days. At other times, you can watch as the conditions track across a portion of the country. One morning, I found that the Chicago area was into Oklahoma City on 70 cm. By late morning, conditions had deteriorated and there were no signals into Oklahoma City. However, to the east in Tulsa, things were just getting started.

Even though Oklahoma City was excluded, for the next three days conditions provided openings between Tulsa and to the east and northeast. While operators in Oklahoma City could hear the Tulsa stations working the DX, they couldn't hear the other end of the QSOs.

Tropospheric propagation can even be tricky in your own neighborhood. A couple of years ago Glenn Bishop, WN5J, was enjoying an opening that extended up to 23 cm. Glenn is located at an elevation of 1,800 feet above sea level (ASL). I'm located about 12 miles away at around 1,000 feet ASL. I couldn't hear the stations Glenn was working because I was below the propagation.

Nevertheless, most of the time tropospheric enhancement operation is a fairly easy way to work those close-in grid locators (and an occasional long-haul DX contact). It requires no special protocol or equipment. Just listen for the presence of conditions and join in.

Scatter Propagation

Six meters is never really dead.

What's that you say? There are no signals on the band; how can the band be open? VHF+ operators discovered long ago that faint propagation exists at all

While on trips author N6CL has been known to use a rental car for an antenna farm. The hood is raised so that there is access to the battery for power to an FT-650. On top of the car is a 6 meter SQLOOP. On the trunk of the car is the 2 meter SQLOOP.

times on these bands. Using scatter propagation, you can work distances up to 1,200 miles on a regular basis. Most of the time signals are just below the noise. However, if you're diligent, you can pick out enough information to make a complete contact with a distant station. While not much is known about this form of propagation, it is used on 6 and 2 meters to make otherwise improbable contacts.

While it's preferable to have a well-equipped station and lots of power for 6 meter scatter work, it is possible to make contacts with just a 150 watt brick. However, you need to be in a quiet location, free from manmade noise sources. You also need a good set of ears and the ability to null out the band noise. Finally, you need lots of patience. It will take some time to complete the contact, snatching bits and pieces of propagation from the ether.

On 2 meters, scatter work takes considerably more power, an even larger antenna, and more patience. Nevertheless, it can be done.

Another form of 6 meter scatter propagation occurs when signals are bounced back from an ionized F_2 layer. Sometimes an F_2 layer is so intense that signals hitting it at just the right angle are refracted back to the source. Operators trying to work South America have sometimes noticed that the signal hits the first F-layer, bounces back to Earth on the Gulf of Mexico or the Caribbean, and then scatters back across the surface of the Earth to its source.

I've heard this form of propagation on the HF bands and 6 meters. The signals are weak and take on a bassy and sometimes watery sound.

Tropo scatter is another form of propagation. Tropospheric scatter differs from tropospheric bending or ducting as described above, because in those forms the signal seems to be held in place as it passes along the path of propagation. With tropo scatter propagation, the signal seems to be bouncing off clouds of ionized air in the troposphere. These signals are described as being weaker than tropo duct or inversion-induced propagation and more watery sounding.

Rain scatter is another type of scatter propagation. Found on 23 cm and above, rain scatter is most noteworthy on frequencies with wavelengths short enough to be affected by the size of raindrops. Experiments conducted on 10 GHz in the rain have shown that each operator, by pointing his antenna at a rain squall, could hear the other station via the signal being propagated by refraction off the raindrops.

Lightning scatter is related to rain scatter, and is a very dangerous and extremely unreliable form of propagation. Operation via this mode relies on the refraction of a signal off the ionized air left by a lightning strike. In order to use this form of propagation, you must be in the middle of a thunderstorm. Operation via lightning scatter is a good way to get into the Silent Key column.

References

1. Ross A. Hull, "Air-Mass Conditions and the Bending of Ultra-High Frequency Waves," *QST*, June 1935, pp. 13–18, 74, and 76.

Chapter 11

Meteor Scatter Communications

Meteor scatter is probably one of the most exotic forms of propagation on the VHF+ frequencies. Here's bit of history and some instructions for making successful meteor scatter contacts.

It was the night of October 9, 1946. There were so many stars, it seemed as if the sky was falling. The Giacobinid-Zinner Comet and its associated meteoroids were making their predicted rendezvous with Earth. What wasn't predicted, however, was the effect the meteor vaporization would have on the VHF+ ham bands.

Amateur radio operators had already experienced bursts of reception from distant stations during meteor showers. This was called the "shooting star" effect. However, until 1946, no definitive work had been written concerning the use of meteor vaporization-induced ionization for propagation.

Then, in January, Oswald Villard, Jr., W6QYT, wrote an article for QST called "Listening in on the Stars."[1] In this article he described how meteor vaporization-induced ionization of the E-layer worked, and what to expect when listening to distant shortwave stations. Villard discussed how the ionized trail would be at a slant in the atmosphere, and how this slant would cause a Doppler shift in frequency of any signal it refracted. He also noted that such a Doppler-shifted signal would heterodyne with the signal being refracted by normal ionization of the atmosphere, causing a whistling sound in the receiver.

The 6 meter ham band had only been assigned for amateur radio use since March 1 of that year. Hams were eager to establish records and see what the band could do in the way of propagation. Because of this curiosity, the familiarity with the "shooting star" effect, the article written by Villard, and the publicity surrounding the forthcoming meteor shower, there was a lot of anticipation about what might occur on that October night. Both amateur astronomers and amateur radio operators were eagerly awaiting the effects of the shower.

What happened on 6 meters was beyond description! In her December 1946 column, Jo Conklin, W9SLG, the original editor of CQ's VHF column (then called "U.H.F."),[2] printed a summary from a letter written by John Taylor, W3OMY. Taylor indicated that " . . . he enjoyed a lot of what look(ed) like short sporadic-E distance or long ground-wave or low-atmosphere-bending DX . . . " He went on to say that there were numerous stations heard in heavy QRM.

In his December 1946 column, Ed Tilton, W1HDQ, then editor of QST's "The World Above 50 Mc.,"[3] commented that the event " . . . produced such bursts, occurring over a wider area and for a longer period of time than ever before in the history of VHF work." Reports sent to Tilton from numerous amateurs indicated that the band opened around 8:30 p.m. He described the sound they heard as a "peculiar rumble" noticeable on all signals coming from beyond the horizon. Tilton related that reports he received indicated that signals from as close as 200 miles and as far as 1,200 miles were coming in all at once, causing QRM that " . . . had never before been experienced on the normally wide open space (of the 6 meter ham band)."

Ed Ladd, W2IDZ, one of the participants that night, described the QSOs as "finished." Not only were signal reports and other necessities exchanged, but unnecessary items such as equipment and weather reports were swapped, as well. Ladd reported that after QSOs were completed, operators continued to hear the other stations for two to three minutes. He described the contacts as fluttery with intermittent drop outs of signal strength as the ionization disappeared and reappeared with each meteor vaporization.

In all, the propagation lasted for around three hours. Most of Tilton's reports came from the eastern and midwestern portions of the United States. For the first time in the history of ham radio, meteor scatter was put to genuine use for signal propagation.

This shower piqued the interest of hams and other communication types in this form of propagation. John Stewart, et al,[4] reported on the results of a U.S. Army radar experiment conducted during the shower that showed echoes on the 106 MHz "early warning" radar. (The shower became known as the *Draconids* because the radiant, the point of origin in the sky of the meteor shower, appeared to radiate from the head of the Draco star constellation.) The article also reported that no evidence of echoes was detected on radars operating on 600, 1,200, 3,000, or 10,000 MHz.

In November 1946, Gurdon Abell, Jr., W2IXK, wrote a letter that appeared in the "Correspondence From Members" column in *QST*.[5] Abell discussed the results of an experiment he'd conducted during the previous August's *Perseids* shower. Using information from the Villard article, he set his HF radio (an SX-25) to hear the whistles of the meteors. He then set his 2 meter super regenerative receiver alongside the SX-25. Curiously, Abell reported that at the same time he heard the whistles on the SX-25, he also heard bursts of signals from distant stations on the 2 meter radio. He tried several variations on this experiment to make sure that he wasn't fooling himself. Abell finally concluded that the bursts of signals must be related to the meteor showers.

Following the October shower and this letter, Mr. Villard wrote to the editor.[6] He commented on Abell's experiments along with experiments of his own conducted during the October shower. Then in July, Villard wrote another article—"Meteor Detection by Amateur Radio"—which ran in the July 1947 issue of *QST*.[7] While not directly acknowledging the veracity of Abell's experiments, Villard did provide some credibility to them by citing them as a reference.

From this bang of a beginning, meteor-shower-induced propagation became a part of the VHF+ operator's communications repertoire. However, it wasn't until 1953 that a 2 meter contact took place using this mode.

In June 1953, Paul Wilson, W4HHK, and Ross, W4AO, were in contact via a tropo path. After the path fell apart, Paul continued to hear signal bursts. Ross advised Paul that these were meteor bursts.

Within a few days of this contact, Paul got a letter from Ralph "Tommy" Thomas, W2UK (the same Tommy Thomas who later became KH6UK and set the tropo records mentioned in Chapter 10), asking to set up schedules for a possible 2 meter contact via any mode of propagation. Paul responded that he'd like to try to work him via meteor scatter.

Over the next several months, schedules were set without success. Then on the morning of October 22nd, it all fell together. Tommy copied more than 2 minutes of transmission from Paul and he, in turn, was able to copy Tommy's confirmation and signal report. With that exchange, they snagged the first complete 2 meter QSO via meteor scatter.

Interest in meteor scatter grew over the years. In April 1957, Walt Bain, W4LTU, wrote for an article for *QST* entitled "V.H.F. Meteor Scatter Propagation." It has become the classic commentary on meteor scatter work. Topics covered included what to expect when attempting a contact, and suggested sequencing lengths. Because CW was used as the mode of communication, there was support for 5 minute sequencing lengths. However, today's almost exclusive use of SSB makes anything but 15 second sequencing archaic.

Interest continued to grow as an ever increasing number of operators tried the mode. Remember that in those days operators didn't have sophisticated equipment with fancy filtering and digital readout. With analog readout, calibration as close as 5 kHz was considered excellent. Imagine trying to find a burst when you're not sure where to look. Also, the primary mode of communication was high speed CW (often around 40 wpm). It wasn't a method of propagation for the faint of heart. Nevertheless, encouraged by intensified publicity more operators tried it out.

A number of operators got excellent results from the *Leonids* shower in 1965. Continued good press prompted more interest. Encouraged by predictions in the November 1966 issues of both *Sky and Telescope*[8] and *Natural History*,[9] hams stood by for what they thought might be a better than average night for the 1966 *Leonids* shower.

And it was! The headline for Sam Harris, W1FZJ's "World Above 50 Mc." column in the January 1967 *QST*[10] was "November Leonids—Shower of a Lifetime." Sam recounted, "Hundreds of contacts were made by calling CQ, or by breaking stations when their skeds were completed, as most were in the first minute or two of prearranged calls."

Reports of visual observations were sent to *Sky and Telescope* from all over the country. One report came from Shelby Ennis, W4WNH (now W8WN). Shelby wrote, "For us in Kentucky, the 1966 *Leonids* will be rated much better as a 'radio' shower than as a 'visual' shower due at least in part to the very sharp peak coming after dawn." However, in areas where dawn hadn't come, particularly in the west, the display was awesome. Reports of 2,000 meteors per minute weren't uncommon. It was a night (or early morning) to remember for amateur radio operators and amateur astronomers, alike.

Making Contacts Via Meteor Scatter

While earlier illustrations indicate that contacts made during previous storms were "nearly normal" in that the operators were able to copy each other for periods of 2 to 3 minutes at a time, contacts attempted via meteor scatter are much different in structure.

During a meteor shower, random contacts are often possible. One station will call a very brief CQ and listen for a response. A station hearing the first station will call that station, give his callsign, and either his grid locator or a signal report. The first station then announces the calling station's callsign and gives the responding grid locator or signal report. The second station responds by saying "Roger" several times. The contact is considered complete if both parties have all they need for the QSO. The entire contact may take as little as 10 seconds to complete, if that.

For most meteor contacts, however, a structured schedule is set between two stations who wish to talk to each other. If you set such a schedule, you'll probably run for half an hour. You'll transmit for 15 seconds and listen for 15 seconds. The westward station transmits first. Some operators break at the end of 7 seconds and listen briefly for the other station. Be sure to clarify operating procedures with the other station before beginning your sked. The initial exchange includes the other station's callsign and your callsign, without either of you saying "this is."

For example, if I, in Oklahoma, grid locator EM15, were running with Jack, AB4CR, in Kentucky, grid locator EM78, I'd start by saying "AB4CR N6CL" over and over again for 15 seconds. I would then listen for Jack to repeat "N6CL AB4CR" over and over again during his 15 seconds of transmission time.

After one of us has heard "complete callsigns," the receiving station starts transmitting a signal report. So, when I've heard both my call and Jack's call (in no particular order), I start repeating "S-2" during my 15 second segment, interspersing our callsigns—just in case Jack has yet to hear complete callsigns.

The signal report of "S-2," rather than the traditional "59," is a way of telling the listener the length of the burns being heard. The letter "S" stands for the word "signal" and the number 1, 2, or 3 stands for the length of the burn. Number 1 stands for "pings"; number 2 stands for burns long enough to make a contact; and number 3 stands for very long burns—at least 15 to 30 seconds in length. Therefore, a signal report of "S-2" means that the sending station is hearing the receiving station on burns long enough to make a contact. As a matter of convenience, most operators stick with "S-2," much like HF operators stick with "59."

Assuming Jack has heard both calls and the signal report "S-2," he'll start saying "Roger, S-2," over and over again during his 15 seconds. Once I have heard "Roger, S-2," I reply with "Roger" over and over again. Once Jack has heard my "Rogers," the QSO is complete. As an option, Jack can come back and say "Roger, 73" repeatedly during his sequence. However, it's not necessary to complete the contact.

Occasionally, the sequence can be broken. For example, when I ran with Ted Goldthorpe, WA4VCC, during the 1992 *Perseids* meteor shower, I heard him give callsigns during the last 3 to 4 seconds of his 15 second segment. I immediately said, "WA4VCC N6CL. WA4VCC N6CL. S-2, S-2, break." Hearing me, Ted came back and said, "Roger, S-2. Roger, S-2, break." Continuing to hear him, I replied, "Roger, roger, roger, break." Hearing my "Rogers," Ted responded, "Roger, 73. Roger, 73. Break." I then replied "73, 73." At that point, we both considered the contact more than complete just 3 minutes into the half-hour schedule.

Band Conditions During A Storm

Occasionally a meteor shower can be so intense that it is considered a storm. What band conditions can you expect during the most intense form of meteor storm? Unfortunately, it's not entirely possible to predict band conditions with certainty, especially considering what propagation modes may be present at that time (sporadic-E, tropo, etc.). However, some generalizations can be made based on past experiences. On 12 meters, it will seem like the band is open everywhere (on short skip) during the peak. On 10 meters, conditions will be much the same. If the storm is very intense, the same conditions that exist on 10 meters may also be present on 6 meters. On 2 meters, stations may have propagation over a given path for up to a

METEOR SHOWERS

Shower Name	Month (and Days)	Peak Date(s)	Best Path(s)
Quadrantids	1-6 January	3-4 January	NE-SW, NW-SE
Lyrids	18-25 April	21-22 April	N-S
Aquarids	21 April-12 May	4-5 May	NE-SW, NW-SE
Arietids	29 May-19 June	7-8 June	N-S
Perseids	23 July-20 August	11-12 August	NE-SW, NW-SE
Orionids	2 October-7 November	20-21 October	N-S
Taurids	20 October-20 November	3-4 November	N-S
Leonids	14-20 November	17-18 November	N-S
Geminids	4-16 December	13-14 December	N-S
Ursids	17-24 December	22-23 December	E-W

Table 11-1. Names, peak dates, and best paths of the more well-known meteor showers.

minute, or so. On 135 cm, propagation may exist for up to 5 seconds, or more. Propagation on 70 cm may exist for a fraction of a second to a couple of seconds.

Meteor Shower Dates

While the *Perseids* meteor shower remains the most popular, others occur almost every month. Also, sporadic meteors—those not associated with a shower—fall all the time, particularly during the early morning hours between July and December.

As stated earlier, meteor showers are given names associated with their position in the sky. A shower appears to originate at a particular point in the sky and spreads out from there. The name assigned to each shower is usually related to the constellation closest to the shower's hypothetical point of origin. For example, the *Quadrantids,* or *Quads* for short, the first shower of the year, is named after Quadrans Muralis, now an extinct constellation. Table 11-1 lists the names, peak dates, and best paths of the more well-known meteor showers.

Popular and Not So Popular Meteor Showers

January: The *Quadrantids* (or *Quads*, for short), a brief (14-hour long), but prolific meteor shower, appears the first week in January. The best paths are north-south. Long duration meteors can be expected about 1¹/₂ hours after the predicted peak. Look for north-south propagation from these, but don't be surprised by propagation from any path.

April: The shower for this month is the *Lyrids*. The best paths are north-south. Although it's a minor shower, on April 21, 1982 at 0650 UTC American astronomers observed a peak that averaged 3 to 5 meteoroids per minute. The spike was over in an hour.

May: The *Eta Aquarids* is a 3-day (between May 4th and 6th) shower. This shower is more popular in the southern latitudes because of the low (−1°) elevation of the radiant. Nevertheless, stations in the southern portion of the United States and stations in Central America can benefit from propagation induced by this shower. While not a productive shower, it did produce in excess of 110 meteoroids per hour in 1980. The best paths for propagation are northeast to southwest and southeast to northwest. East to west propagation is fair and north to south propagation is poor.

Additional potential for meteor scatter contacts may exist at the end of the month as the first signs of the *Arietids* begin to make their appearances. The minor showers of *Herculids* (May 19 to June 14, peak June 3 to 4) may also have some effect on elevating the number of meteors. The *Scorpiids* (May 29 to June 20, peak June 5 to 6) is a southern latitude shower. Stations in the southern hemisphere may benefit from the slightly (10 meteoroids per hour) elevated number of meteors entering the atmosphere.

June: Between May 29 and June 19, the *Arietids* meteor shower makes its annual reappearance. But don't look for the meteoroids in the sky; this is principally a daytime shower. The north-to-south path seems to be the best for this shower, with the peak days being June 7–8.

August: The most popular shower of them all is the *Perseids*, which peaks around August 11 and 12. This shower is popular because it occurs in the summertime and because it's a long-lasting shower. Meteoroids from the *Perseids* start appearing as early as July 15. However, after August 13, the shower is all but over.

September: On September 1st the α *Aurigids* meteor shower may peak. This is a minor shower that showed some life in 1986 (a 1 hour shower was observed in Hungary). Its radiant position is around 42°, in the constellation α *Aurigae*.

October: The *Draconids* shower has a checkered past. In 1985, it produced 200 meteoroids per hour at its peak. However, since then there has been insufficient activity to support much meteor scatter communications. Nevertheless, operators reported a marked increase in activity during the 1993 shower, particularly between 1230 and 1400 UTC on October 10. Normal peak days are around October 9 and 10. The next possible peak of shower activity won't occur until 1998.

The peak of the *Orionids* shower comes around October 22 and 23. It's considered a minor shower, producing only 8 meteors per hour at its maximum. While visibly entertaining, it's not much use for meteor scatter communications.

While neither of the October showers show much promise, it's not out of the question that something will happen in space during the months leading up to their scheduled appearances. Effects of the Sun's gravitational pull on comets have created unexpected showers and storms. As always, being prepared is the best bet for success in any VHF+ communications.

November: The showers for this month include the *Taurids* and the *Leonids*. The peak day for the *Taurids* is November 4 or 5. The peak day for the *Leonids* is November 17 or 18. Neither of these are normally strong showers. However, the *Leonids* have been known to produce storms, and it is predicted that one may occur in 1998 and possibly 1999.

December: There are two showers scheduled for this month. The first, the *Geminids*, tends to peak December 13. It's a good north-south shower, producing an average of 50 to 60 meteoroids per hour at its peak. The second, the *Ursids*, usually peaks around December 22. It's an east-west shower, producing an average of only 10 meteoroids per hour at its peak.

References

1. Oswald Villard, Jr., W6QYT, "Listening to the Stars," *QST*, January 1947, pp. 159–60, 120, 122.

2. Jo Conklin, W9SLG, "U.H.F.," *CQ*, December 1946, p. 34.

3. Ed Tilton, W1HDQ, "The World Above 50 Mc," *QST*, December 1946, p. 43.

4. John Stewart, et al., *Sky and Telescope*, issue 65, pp. 3–4.

5. Gurdon Abell, Jr., W2IXK, "Correspondence from Members—," *QST*, November 1946.

6. Oswald Villard, Jr., W6QYT, "Correspondence from Members—," *QST*, January 1947, p. 61.

7. Oswald Villard, Jr., "Meteor Detection by Amateur Radio," *QST*, July 1947, pp. 13–18.

8. *Sky and Telescope*, November 1966, p. 251.

9. *Natural History*, November 1966, pp. 43–44.

10. Sam Harris, W1FZJ, "The World Above 50 Mc," *QST*, January 1967.

Chapter 12

Auroral Communications

If you've spent any time on HF, you're somewhat familiar with the effects of aurora, particularly on 10 meters. Those watery sounding signals you've heard are signals affected by aurora. However, that's only one small aspect of auroral propagation.

In a *QST* article entitled "Radio Aurora,"[1] Richard Miller, VE3CIE, states that aurorae are caused indirectly by solar flares and coronal holes. Either one of these activities can cause particles to be spewed out from the Sun's surface. These particles travel in the solar wind and are caught by the Earth's magnetic fields. The resulting ionization affects the D, E, and F propagation layers of the ionosphere. Whereas the D and F layers are adversely affected, each in different ways, the E layer is affected positively—at least for the VHF+ operator.

When the D layer is saturated by ionization caused by these particles, the HF bands may suffer from "blackout." The higher the ionization of the D layer, the higher the frequency of the blackout.

Across the polar path, the F layer may be adversely affected to the degree that signals sound weak and watery, or may not get through at all. However, the affects of ionization on the F layer may not be all bad, particularly across the equator. Intense ionization of the F layer can create propagation across the equator on the 6 meter band (and possibly the 10 meter band). When equatorial propagation conditions exist, operators, and particularly those operators in the southern states, can work stations in the northern parts of South America.

On the higher VHF+ frequencies, aurora is often observed as a buzz on the signal. To the uninitiated, the sound is that of a transmitter with no filtering in the power supply. However, even that comparison isn't quite accurate, because the sound induced by a lack of filtering is often either a 60 or 120 Hz tone superimposed on the signal.

The buzz of the aurora signal is random. This is because the signals are refracted off different surfaces of the aurora oval all at once, causing Doppler shifts in the frequency of the signal. These Doppler shifts are the combination of both the shift in the center frequency and the broadening of the signal.

Center frequency shifting can be so severe at times that you'll have to tune your radio to follow it across the band. The flattening of the signal can be so severe that it becomes indistinguishable from noise. Also, the higher the frequency, the more pronounced these Doppler shifts become. In fact, Harry, K3HZO, and Paul, WA3NZL, found the Doppler shift on their record-setting 33 cm contact was between –2.5 and –3 kHz.

Due to the buzz, contacts are almost always made on CW because SSB is unintelligible. Occasionally, though, the buzz will disappear and the signals will become clear and loud. This effect is the result of auroral-E propagation. Not much is known about auroral-E except that occasionally, after an auroral event has been in progress for some time, the E layer becomes so ionized that it supports signals much as it would during sporadic-E. Auroral-E is most often observed on 6 meters, but has made appearances on 2 meters. Bill Tynan, W3XO, speculated in his column[2] when reporting on the February 8, 1986 aurora, that as intense as the auroral-E propagation was on 2 meters, it might have supported signals on 125 cm.

When Can Auroral Propagation Be Expected?

There are certain times of the day that are best for aurora. The peak time seems to be around 6 p.m. local time, with a secondary peak sometime around 2 a.m. During the year, the best times seem to be around the spring and fall equinoxes. Also, look for aurora 27 days after a

As stated by VE3CIE, aurorae are caused indirectly by solar flares and coronal holes. The resulting ionization affects the D and F layers of the ionosphere adversely, and the E layer positively—at least for the VHF+ operator. (Photo courtesy Dennis di Cicco and Sky and Telescope)

major solar event has occurred. If the disturbances have survived the rotation of the Sun, it's possible they may cause another event one rotation later.

For those of you who plot the Sun's activities, it's important to note that there's an increase in these disturbances somewhere between 2 to 4 years after the peak of the solar cycle.

Because these events are related to the Sun's activities, it's possible to make some general observations. Solar indices are broadcast on radio station WWV on 5, 10, 15, and 20 MHz at 18 minutes past the hour. Of the three indices, the A- and the K-index are particularly noteworthy because both indicate relative disturbances on the Sun. In particular, if the K-index is a 6 or higher, there's a strong likelihood that an auroral event will occur. If the K-index is a 9 (as it was twice during the February 8, 1986 aurora, and for an extended period during the March 13–14, 1989 aurora), get on the air immediately!

Another item to listen for during WWV broadcasts is the announcement of a solar flare and a subsequent proton event. Should this occur, you may be experiencing some excitement on the VHF+ frequencies from one to two days later.

How To Work The Aurora

Because you need to bounce your signal off the auro-

ral curtain, you must point your antenna at the curtain. It may be directly north, or it may be off to one side or the other. It's possible to improve your ability to bounce your signal off the curtain by elevating your antenna, so it helps to have an elevation rotator along with the azimuth rotator. Because the auroral curtain is moving constantly, there's no clear way of determining where the best path will be. Therefore, while you're spinning your dial looking for stations to work and tracking them up and down the band, you'll also need to rotate your antenna back and forth and, to a small degree, up and down.

While these techniques will help you work the auroral curtain, they won't work for auroral-E. Should auroral-E propagation appear, you'll find that the paths between you and the distant station become more direct. Although it may take a bit of getting used to, you'll find you can work stations in excess of 1,000 miles via auroral propagation on 2 meters through 70 cm, and even farther via auroral-E.

References

1. Richard Miller, VE3CIE, "Radio Aurora," *QST*, January 1985, pp. 14–18.

2. Bill Tynan, W3XO, "The World Above 50 MHz," *QST*, May 1986, pp. 68–69.

Chapter 13

F₂ and TE Communications

Being rarer forms of propagation that can affect the VHF+ frequencies, F₂ and TE are both discussed here because they influence these bands less frequently.

F₂ Propagation

The F layer is the atmosphere's highest layer, and is found between 100 and 300 miles above the surface of the Earth. During the peak of the sunspot cycle this layer receives ionization that will support refraction of wavelengths into the 6 meter band. Worldwide propagation is possible during the years surrounding the peak of the sunspot cycle.

Six meter enthusiasts plan for these times in order to complete the necessary contacts for achieving DXCC. In fact, the peak of this cycle (cycle 22) produced the first recipients of this award and boosted many others within tantalizingly close proximity of their goal. No doubt those operators who didn't quite make it during cycle 22 will do so during the peak of cycle 23.

When the solar cycle is at its minimum, little F layer propagation occurs. Many operators actually disassemble their 6 meter stations and store them until the peak of the next sunspot cycle.

What can you expect from the next sunspot cycle? Peter Taylor, in his book *Observing the Sun*,[1] examines recent cycles and compares the even-numbered with the odd-numbered cycles. Taylor concludes that even-numbered cycles have longer extended maxima than odd-numbered cycles. However, he also points out that recent odd-numbered cycles have been higher than their counterpart even-numbered cycles. Finally, he notes " . . . if the odd-even relationship continues, the maximum of cycle 23 should also be a very strong one, perhaps with a peak strength which approaches 200 (in mean sunspot numbers)."

What About F₂ On Other VHF+ Frequencies?

The MUF of F layer propagation rarely reaches 70 MHz. Only on very rare occasions have European amateurs, who have the 70 MHz ham band, made contact with stations via this form of propagation. In fact, the 6 meter ham band was actually an FCC compromise that recognized the rarity of this form of propagation. Before World War II, amateurs in the United States had the use of the 5 meter band, which extended between 56 and 60 MHz. As a way of allocating frequencies for the new television services, the FCC set aside certain blocks of 6 MHz for the lower channels. Originally, the Commission was going to give the amateurs a band between 44 and 48 MHz. However, intense lobbying by the ARRL convinced the FCC that there was sufficient occurrence of both sporadic-E and F₂ propagation such that the band " . . . would possess small novelty and much of the eager interest of amateur observers would disappear."[2] Owing to the League's urging, the FCC created a channel 1 beginning at 44 MHz, and then granted amateurs the 6 meter band between 50 and 54 MHz. The allocations continued with the assignment of channel 2 between 54 and 60 MHz. (Eventually, the FCC abandoned the channel 1 assignment and later subdivided it for use by fixed and mobile services. Such services serve as beacons today, alerting 6 meter operators to possible impending openings on their band.)

While most F₂ propagation disappears during sunspot lulls, some signals are occasionally disseminated by this mode. No one knows why; it just happens!

TE Propagation

Transequatorial, or TE, propagation is related to F_2 propagation in that its signal is refracted by the F layer. TE also seems to occur most often during the peak of the sunspot cycle. Additionally, TE propagation seems to occur more often in the spring, during the late afternoon or evening.

To take advantage of TE propagation, both you and the station you're trying to work must each be the same distance from the equator. The range of distance is about 2500 miles either side of the equator. Unfortunately, this rules out all but the southern tips of Florida and Texas and the southern west coast on the continental United States. Nevertheless, it does include stations on the opposite end of South America, southern Africa, and in the Pacific.

Although it has yet to be reported, propagation up to 432 MHz is possible. With sporadic-E link-ups, occasional contacts to more northern QTHs on the continent can occur on 6 meters. More rare are meteor burst links with TE propagation. One such event is believed to be the cause of the contact that Larry Lambert, NØLL, had with Nob, VR6JJ. Larry reported that he could barely hear Nob, until all of a sudden he burst through. They quickly completed the contact, and then Nob was gone. Larry attributes that sudden burst to ionization caused by a meteor burn.

How Does TE Propagation Work?

Most of the time the southbound signal travels outward to an F_2 layer north of the equator, is refracted back to Earth at the equator, bounces outward to another F_2 layer south of the equator, and is finally refracted back to Earth. However, sometimes these two layers break up into ionized clouds and traverse the equator. When this happens, the signal appears to become trapped below these clouds and is continuously refracted until it lands on the surface at the distant location. It is this breakup, which seems to be what occurs during an auroral event, that creates the transequatorial opening on 6 meters.

References

1. Peter O. Taylor, *Observing the Sun*, Cambridge University Press, 1991. It's interesting to note that indirect correlation to this prediction existed as far back as 1976. In a phone conversation I had with Dr. John A. (Jack) Eddy, the subject of my July 1976 *QST* article on the Maunder Minimum, he expressed the feeling that we were headed for another Grand Maximum of a long-term solar cycle that stretches into 200 to 300 years in periodicity and that this maximum would probably occur within the 21st century.

2. Excerpt from a Brief that appeared in August 1945 *QST*, p. 12.

Chapter 14

EME Communications

During the month of October, there's an annual renewal of interest in moonbounce, or Earth-Moon-Earth (EME), communication. Although EME has been around since World War II, successful day-to-day amateur radio communication using this mode is relatively new. The first complete amateur two-way EME communications didn't occur until 1960, and it wasn't until the advent of the United States higher power limit of 1500 watts output and the arrival of GaAsFET preamps in the 1980s that EME communication became more popular.

Among the VHF/UHF bands, 144 MHz is the most popular for EME communication. Although EME communication has been successful on 50 MHz, the size of the antenna arrays and background sky noise restrictions remain barriers for all but the most serious operators on that band. The higher the frequency, the higher the path loss; therefore, more elaborate antenna arrays are required for successful EME work above 144 MHz. Accordingly, most operators start on 144 MHz and, if they find EME is for them, try the higher frequencies later.

To gather information on this aspect of VHF+ DXing, I talked to several 144 MHz EME communication experts: Bev Cavender, W4ZD, San Hutson, K5YY, John Carter, KØIFL, and the dean of 2 meter EME, Dave Blaschke, W5UN. What follows is a compilation of their thoughts about EME communication on 2 meters.

Several factors affect EME communication. These include libration fading, tremendous path loss, noise (both Sun and background sky), Faraday rotation, and spatial polarization.

Because the Earth and the Moon wobble along in their orbits, signals emitted from Earth stations hit a target *area* on the Moon, rather than a bull's eye.

Also, because the Moon's surface is very irregular, the reflected signal takes on that irregular shape. The combined effects of the wobbling orbits and irregularly shaped signals cause fading and a certain amount of Doppler shift between stations attempting communications. This is called libration fading. When operating on 2 meters, you'll experience longer term peaks and valleys, where portions of a callsign will be heard clearly, followed by very weak signals. While these effects aren't nearly as pronounced on 144 MHz, the effect on 1296 MHz may be as high as 20 dB fading and 10 Hz frequency shift.

The Moon is located over 221,000 miles from Earth at perigee (the closest point to Earth) and over 252,000 miles from Earth at apogee (the farthest point from the Earth). Due to the shape and size of the Moon, only about 7 percent of the signal that strikes it is reflected. The remaining 93 percent is absorbed or misses the Moon and is lost for communication. The path loss is directly proportional to the frequency of operation—that is, the higher the frequency, the higher the path loss. Therefore, the path loss is around 252 dB at perigee and 254 dB at apogee on 144 MHz. For the low power station, the 2 dB difference between perigee and apogee may be just enough for a successful QSO.

Noise, caused by the Sun and the background sky, inhibits your ability to receive weak signals. For those of us in the northern hemisphere, communications generally aren't favorable the day of a new moon (you won't be able to see the Moon with the naked eye, except in an eclipse) or at times when the Moon is farther south in the sky. Communications are less favorable during times when the Moon travels more to the south, not only because of increased background sky noise caused by constellations in the southern sky, but also, Dave says, because of convention. The higher latitude European stations see less of the Moon when it's farther south; consequently, they don't get on the air. The most ideal time of the month for northern hemisphere stations tends to be when the Moon has finished its most northerly declination and is moving southward in the sky.

Faraday rotation is the polarization rotation of a signal because of the influence of the Earth's ionosphere on that signal. Some say this is the result of the effect of the Earth's magnetic field on the signal as it passes through the ionosphere. (Dave has noticed some correlation between what happens with Faraday rotation and what happens on HF propagation. It remains one of the mysteries of EME communication and deserves further study.) Faraday rotation affects the signal by causing it to go through a deep cyclical fade. This cycle changes in period, from shorter to longer, as the frequency is increased. It is estimated to have a period of approximately 20 minutes on 144 MHz. Dave says the cycle is more pronounced on some days than on others. QSO schedules are set up to accommodate this period. These schedules last typically for $1/2$ hour to 1 hour on 144 MHz, with 1 hour for casual schedules and $1/2$ hour for contest schedules. Although some contest QSOs operate on schedules (particularly low power stations wanting to work high power stations), most contacts are random.

Spatial polarization simply means that two stations at different locations on the Earth are aiming antennas fixed in the (horizontal or vertical) plane at the Moon. Using a mirror analogy, if you were to look at something at an angle with a mirror, depending on how your head is tilted, that object may appear right side up, at an angle, or upside down. If one of the stations has the ability to rotate the antennas through the plane between horizontal and vertical, some of the effects of spatial polarization can be overcome. However, rotating several antennas through this plane simultaneously, while maintaining phasing relationships between each antenna, becomes a bit of a mechanical nightmare. Therefore, spatial rotation is often overcome by brute force. Adding more and more elements to an antenna array helps reduce the effects by increasing the array's dB gain. Also, Dave notes that Faraday rotation has a tendency to overcome spatial polarization during at least part of the scheduled period for a QSO on 2 meters.

There are two other points to keep in mind concerning EME communication. First, on moonrise, you'll experience Doppler shift of between 300 and 500 Hz above your frequency. On moonset, the Doppler shift will be 300 to 500 Hz below your frequency. When the Moon is overhead, there is no Doppler shift. Those of you who have worked the satellites are familiar with the effects of Doppler shift and keep your hand on the tuning knob. Second, if you are able to hear your echoes, be prepared for a 2.3 to 2.7 second delay. That Moon is a long way off, and it takes time for your signal to get there and back.

CW is the preferred mode of communication on EME. It's the most reliable mode due to the weakness of the signal. The transmission is at a rate between 10 and 15 wpm. Slower CW can break up as a result of fading and fluttering, while letters transmitted using faster CW tend to disappear.

EME communication is similar to meteor scatter in one sense: both are dealing with weak and irregular signals. Therefore, as with meteor scatter, EME communication has a protocol. However, because of the nature of the EME signal, the procedure is very different from the protocol used for meteor scatter.

The preferred frequency of operation for schedules is above 144.030. The preferred frequency of operation for random QSOs is between 144.000 and 144.030 MHz. If signals are loud enough to sustain SSB QSOs, the preferred frequency is around 144.150 MHz and up.

There are some nets you can listen to for information on conditions and schedules. One net coordinates 144 MHz EME communication. It's hosted by VE7BQH and meets every Saturday and Sunday on 14.345 MHz at 1700 UTC, or as soon as the 432 MHz net is finished. Every Monday at 0230 UTC (Sunday evening local time) at 3.818 MHz (plus or minus QRM), a VHF/UHF clearinghouse net meets to exchange information and set skeds. At 9 p.m. EST Monday another VHF/UHF clearinghouse net meets on 3.843 MHz for the same purpose.

Now, let's look at a sample QSO. A sked is set between DL8DAT in Germany and N6CW in San Diego. The QSO is scheduled to last an hour and will start at 0000 UTC. The eastern station (relative to its position on earth) transmits first. In this case, it's DL8DAT. The transmission will last for two minutes. DL8DAT will send the receiving station's call followed by his own as follows: N6CW de DL8DAT, N6CW de DL8DAT, etc. At 0002 UTC N6CW begins an identical routine, sending DL8DAT de N6CW, DL8DAT de N6CW, etc. The two hams transmit back and forth every 2 minutes, until one station hears the other sending complete callsigns.

Once the receiving station copies complete callsigns, he starts the next phase of the sequence. He sends callsigns, as before, for the first 90 seconds of the 2 minute sequence. But during the last 30 seconds, he adds a signal report—the letter "O."

The signal report was once a "T," an "M," or an "O." A "T" meant that the callsigns were just barely detectable. An "M" meant that portions of a call were copied. An "O" meant that complete callsigns were received. However, because the receiving station is looking for complete callsigns, any other report would be a waste of time in completion of the QSO. As a result, the signal report convention has evolved into the letter "O."

Let's assume that N6CW was successful in copying the callsigns and has initiated the second phase of the protocol. It's now up to DL8DAT to hear the signal report portion of the QSO (assuming he's already heard the complete callsign exchange). Once he hears the signal report, he sends "RO" throughout his entire 2 minute time period. This tells N6CW that DL8DAT has heard the signal report (the "R") and is sending a signal report of his own (the "O"). If his country requires him to sign his callsign at the end of every transmission, he sends N6CW de DL8DAT once at the end of the 2 minutes. Otherwise, no callsigns are sent.

When N6CW finally hears "RO," he sends only the letter "R" during his next 2 minute transmission. When DL8DAT finally hears the "R," he sends "73" or "73/SK" during his next 2 minute session—followed by complete callsigns at the end of the transmission (to comply with government rules pertaining to station identification). The QSO is considered complete when DL8DAT hears the "R" sent by N6CW. The honor system comes into effect here, because you are the only one who knows what you heard.

EME and QRP

What does it take to get "on the Moon"? San Hutson, K5YY, was able to complete his WAS, work 32 countries, and add to his grid locator total by spending just $200 more than his initial outlay for his 2 meter station. He has an excellent write-up in the 1990 Central States VHF Society *Proceedings* (available from the ARRL for $12, plus $3.50 shipping and handling; check for most current price before ordering).

Ray Soifer, W2RS, has made over 20 contacts running only 150 watts and a single Cushcraft long boom beam. He presented a very informative paper entitled "QRP EME on 144 MHz: How and Why" at the 1992

This is a fish-eye-lens view of Paul Chominski, SMØPYP, and his 7.6 meter dish. You can look for him on your favorite EME band. (Photo courtesy SMØJHF)

Central States VHF Society conference. His paper is part of the *Proceedings* for that year, which is also available from the League for $12, plus $3.50 shipping and handling (again, check for the current price). Both Hutson and Soifer's write-ups unlock some of the mystique of EME operation for the "little guy."

EME on Other VHF+ Frequencies

While EME has taken place on 135 cm, uncertainty has caused interest to wane in recent years. Now, however, because of the new FCC regulations that set aside a portion of the band for weak signal (in the FCC's words, experimental) work, interest is picking up again. It remains to be seen, however, just how popular EME communications on 135 cm will become.

Seventy cm is perhaps the second most popular band for EME work. It is both easier and harder to get on this band than on 2 meters. Assembling the right antenna array is one of the easier tasks. Steve Powlishen, K1FO, in the second part of his two-part article in *Communications Quarterly*,[1] reported that a four-antenna array for 70 cm is typically 5 feet by 6½ feet, whereas a typical array for 2 meters is 10 feet by 13½ feet. Also, because of the higher frequency, the

Gerald Williamson, K5GW, has returned to 2 meter EME in a big way with this 480-element array. He can usually be found on your moon rise (or set) around 144.009 MHz. (Photo courtesy WA5VJB)

70 cm antennas are much shorter for the same number of elements.

Signal propagation is also a bit easier on 70 cm. While it still takes high power to make it to the Moon, factors described for 2 meters—such as Faraday rotation and sky noise—have far less influence on 70 cm. Here, again, the antenna becomes a consideration. Because the array used for this band is smaller, it's more practical to design polarization rotation into the antenna. This will help overcome Faraday rotation and correct for cross polarization problems encountered when working a distant station.

However, as I said, there are some barriers to working 70 cm. While transceivers are available for this band, serious EME operators generally opt for transverters and sophisticated HF radios. Also, while the antenna construction is easier, feeding it is not. Because of feedline losses found in coaxial cables, hardline is often used. Also, you must use the correct low-loss splitters for feeding multiple Yagis in the array.

While there is some EME activity on 33 cm, the next most popular band is 23 cm. Here the antenna of choice is the dish. With a circularly polarized feed, antenna cross polarization and Faraday rotation almost become imperceptible. Additionally, sky noise is even less of a factor on this band than it is on 70 cm.

Above 23 cm, most EME is experimental. Only a few operators operate regularly on 13 cm; fewer still operate on 9, 5, or 3 cm. While conditions are such that Faraday rotation and sky noise cease to be problems, other challenges crop up. Equipment availability is the chief difficulty. Learning how to operate with

Doppler shift that takes place over tens of kilohertz is another. If you're interested in pursuing these higher bands, you need to work with the experts. Paul Wilson, W4HHK, has one of the best stations in the United States on 13 cm. Jim Vogler, WA7CJO, is the leader on 23 cm.

It's important to note that sequencing on these higher frequencies is a bit different. Rather than lasting 2 minutes, your transmissions will be 2½ minutes long. The last half minute is either reserved for signal reports or nothing, depending on what you've heard from the other station.

Signal reports are also different. While the letters "T," "M," and "O" are the same, their meanings are a bit different. "T" means "I can hear something," "M" means "I have picked up fragments of callsigns," and "O" means "I have copied complete callsigns." While an "M" is sufficient on 2 meters, an "O" is required on 135 cm and above.

This said, there is an exception to these differing procedures. Operators on 135 cm tend to use either the 2 meter or the 70 cm routines—depending on their background. Those who have operated more on 2 meter EME tend to stick with that method, while those who operate on 70 cm prefer that method. Consequently, when you set a sked on 135 cm, make sure you and the other operator agree on the method of sequencing.

Romancing the Moon

Probably the most successful EME DXpedition, in terms of numbers of stations worked during a specific period, was VE3ONT's operation during the 1993 ARRL EME Contest. VE3ONT used the 150 foot dish at the Algonquin Observatory in Ontario, Canada. The Canadian team operating the dish managed to amass a previously unheard of approximate score of 6.5 million points during the two weeks of the contest. What follows is a summary of the planning and actual operation of the station.

When I was a teenager, my mentor, K6BTO, told me about the first ham radio VHF contacts being made by bouncing signals off the Moon. After hearing those stories, I'd go outside and look at the Moon and wonder how it was possible to send that far into space a signal that would be strong enough to be reflected back to Earth. I also wondered if I would ever have the equipment to bounce a signal off the Moon.

This desire to "romance the Moon" infected many other VHF+ operators at that time. Among them was a group of Canadians. One of their members, Peter Shilton, VE3VD, would look at the giant Algonquin Radio Telescope and, in his mind's eye, wonder about the possibilities of using it to bounce signals off the Moon.

Sharing his dream was another member of the group, Dennis Mungham, VE3ASO. Dennis put his wishes into action. Several years ago Dennis contacted the National Research Council, then the managers of the telescope, concerning the possibility of operating a ham radio station from the telescope facility. They were not interested.

However, in 1991 the Institute for Space and Terrestrial Science became the new trustee for the site. Upon hearing that the Institute had as its mission exposure of space communications to Canada's youth, Dennis saw an opportunity to expose the Institute to amateur radio and show its leaders the potential of blending its goals with that of ham radio.

Normally, the Institute charges clients $1,000 per hour for use of the dish. However, knowing that the hams had knowledge and skills, particularly in development of very low noise figure preamps that would perform satisfactorily in the GHz range, Dennis pitched them a tradeoff. The Institute said it might be interested, offering to waive its fees in exchange for services rendered by the hams. Dennis was asked to make a presentation to the Institute's leadership.

Dennis and Peter got together and made their pitch, which included suggestions on how the hams could be of assistance in the Institute's Space Seminars, held on the Algonquin Park campus. The Institute's director, Dr. Wayne Cannon, was keenly interested in what the hams proposed. He asked that the team, which now included Hans Peters, VE3CRU, and Mike Owen, W9IP, write up a proposal. The team prepared a 40-page paper that outlined all aspects of the operation, such as equipment type, location, and safety considerations. A couple of weeks following submission of the paper, the team received the go-ahead to operate radio telescope during the second weekend of the 1992 EME contest.

The 6 meter EME array of Jim Treybig, W6JKV, is shown here pointing straight up. The 16 vertical poles are the 16 six-element M^2 antennas that make up the 96-element array.

Unfortunately, Mother Nature (mother of Mr. Murphy) had other plans. Winds in excess of 60 mph toppled trees everywhere the day before the contest. Some of these trees fell into the power lines leading to the Algonquin site. Undaunted by the initial reports, the team proceeded to the site anyway.

Upon their arrival, they were informed that the only power was a large diesel generator. They were also advised by the site manager that he would not give them permission to move the dish for fear there might be back EMF from the motors affecting the generator. Nevertheless, they charged ahead, mounting equipment and getting organized to operate the dish that weekend should the commercial power return.

As bad luck would have it, they weren't able to operate during the EME contest that year. Nevertheless, the "dry run" helped them get past the learning curve, and gave them ideas on how to improve the assembly and installation operations for the next opportunity to enter the contest.

Because of an accident with some equipment at the site, a proposed trip in the spring was canceled. No effort was made to use the dish during the summer, because various members of the team like to get together at their regular sites for the summer contests.

The next EME contest finally provided an opportunity to operate the dish. Dennis was able to secure permission from the Institute to operate both weekends of the contest. Plans were made and the list of team members grew. It now included Don, VE2DFO, Dennis, VE3ASO, Bob, VE3BFM, Hans, VE3CRU, Dana, VE3DSS, Heather, VE3EMS, Kevin, VE3KDH, Peter, VE3VD, Michael, W9IP, and Craig Morton.

Equipment was assembled in garages. Bob Morton, VE3BFM, Neil, VE3SST, and Dana Shtun, VE3DSS, both at Sinclabs Amateur Radio Products, worked on the antennas. Tommy Henderson, WD5AGO, built and tested the preamps for 432 and 1296 MHz. The amplifiers were supplied by Peter (2 meters), Hans (70 cm), and Dennis (23 cm).

Departure day finally arrived. The second weekend in October was Canadian Thanksgiving weekend, and the nine guys and one gal (Heather, VE3EMS, wife of Peter, VE3VD) left families behind to make the trip. Everyone completed the trek to the park (which included a journey through 75 miles of wilderness) without incident. Owing to their planning from the previous year, assembly and setup went smoothly. All the team had to do was wait for moonrise plus 9 degrees (because the dish couldn't be lowered to anything less than 9 degrees above the horizon). Thanks to Michael Owen, W9IP's RealTrack software, they knew precisely when this would occur. Nevertheless, they tested the equipment before the Moon was in full view of the dish, and were sweetly surprised as to how *loud* (at times S9 + 20 dB) their echoes sounded.

Once the Team VE3ONT contest effort was underway, they found the pileup as awesome as anything they'd ever heard on HF. In fact, it was so big that the decision they had made in the planning stages to work "HF contest style" proved to be the right one. Working HF contest style gave them the improved efficiency of working each station (over that of the conventional sequencing method). In fact, their QSO rate approached that of an HF contest station.

Still, Murphy wouldn't leave them alone. A nasty hum on 432 MHz kept the operators from hearing the really weak stations (the ones they really wanted to work). Nevertheless, the team pressed on through their lunar window, finishing the first day with 246 QSOs and 41 multipliers.

The next day, as planned, they switched to 144 MHz. Again, Murphy stepped in and created problems. The first was a tripped circuit breaker in the feed cabin, which forced them to awaken Kevin (the only one at that time who knew how to work the cherry picker allowing them to reach the cabin) out of a sound sleep, to ride it up the 90 feet necessary to reach the feed cabin and reset the errant circuit breaker. Soon after that Kevin taught the others how to use the cherry picker. The next problem came via a solar event that produced an aurora which degraded conditions for EME work on 144 MHz. Despite the diminished conditions, they were able to make 235 QSOs with 46 multipliers during this 24 hour period.

Far fewer team members were present for the second weekend of the contest. Included on this trip were VE3s ASO, CRU, and VD, and W9IP. As on the first day, they operated 432 MHz. Again, Mr. Murphy showed up. During the night the temperature dropped from 6°C to –10°C in just a few hours. Because of this drop, the 432 MHz amplifier became untuned, causing an extremely chirpy signal.

This time Michael Owen, W9IP, was roused from bed. He made one trip up in the cherry picker, went into the feed cabin, and retuned the amplifier. Upon returning to the control room they discovered that they now had no power out. Michael returned to the feed cabin, this time to replace a bad section of coax connected to the watt meter.

During this run on 432 MHz, the team made an

additional 79 contacts in 4 more multipliers (including yours truly).

The next day they switched to 1296 MHz, where they literally "worked the band dry." For the most part, Murphy left them alone. Their total was 79 QSOs and 29 multipliers.

Overall, after subtracting the duplicate contacts, their score was 560 QSOs in 116 multipliers, for a total of 6.496 million points—obviously the highest total for any station in any previous EME contest.

Speaking of "high," the team was high for several days following the contest. Peter remarked that, for him, the trip was doubly special. First, he had fulfilled his teenage dream of "romancing the moon" far beyond his expectations. Second, having Heather, his wife, along for the first weekend and nearly completing a contact with her during the second weekend meant a lot to him. Peter told me that Heather came to a new understanding of just how important his hobby was to him, and let him know how she had grown to appreciate its importance. For them, it was one of those unique events that only "in love" couples can experience—especially when "romancing the Moon."

References

1. Steve Powlishen, K1FO, "432 MHz EME 1990s Style," Part 2, *Communications Quarterly*, Fall 1991, pp. 33–48. Part 1 of this article can be found in the Premier Issue of *Communications Quarterly*, pp. 29–39.

Chapter 15

What Is A Grid Locator?

If you operate on SSB, CW, or satellite, it's tough to get on the VHF+ frequencies without understanding the grid locator system. Read on for an explanation of what they are, where they came from, and why they are so popular today.

More than 40 years ago a system of grid locators was introduced in Germany as a way of spurring activity on the VHF+ ham bands. These locators were assigned two-letter designators. Initially, the system worked well enough for the areas it covered in Europe and North Africa. However, worldwide expansion of the system necessitated replication of the same two-letter designators in other geographic areas, causing obvious confusion.

Two hams, working independently of each other to alleviate this problem, developed nearly identical designator systems. The first was created by Folke Rosvall, SM5AGM, in October 1979. The system started at the principal dateline and involved 20° by 10° large units, 2° by 1° middle units, and 6' by 3' small units (the measurement is in minutes, not feet).

The second, developed by Dr. John Morris, G4ANB, in December 1979, also involved 20° by 10° large units and 2° by 1° middle units. However, the small units were 5' by 2.5'. The proposed starting location for his system was the Greenwich longitude.

In April of the following year, a group of European VHFers met in Maidenhead, England. Among the 20 or so proposals presented, Rosvall's and Morris's surfaced as the front runners. The group determined that the best solution would be to modify Morris's system to start at the principal dateline.

Grid Locators and The U.S. Awards System

Meanwhile, the activity stimulated by use of the grid locator system in Europe prompted hams in the

B. V. Soerensen, OZ1UM, operated from Spodsberg (JO55wx) several years ago. At that time they set a European record of 90 km for 47 GHz with a QSO between them and DB6NT, DF9LN, and DF2CA operating from Trehoje Mols (JO56gc).

Chip Taylor, W1AIM, operated from his van as VY2QST on a narrow 1/2 by 2 mile strip of land in grid square FN87 (North Cape, Prince Edward Island) on 6 meters in a recent CQ WW VHF WPX Contest. (Photo courtesy W1AIM)

United States to take a look at developing a system for North America. At the 1981 Central States VHF Society conference held in Sioux Falls, South Dakota, the Committee on Society Awards (headed by Lance Collister, WA1JXN) proposed a series of three awards. The first was for making one-hundred contacts on VHF, the second was for making contacts in one-hundred 1° by 1° grid locators, and the third was for scoring 1,000 points by working stations at increasing distances from one's home QTH. Distances were measured on the basis of 1° by 1° grid locators. The proposal was adopted, and the awards were put in place and publicized. In the months that followed activity on VHF+ increased, and a few awards were issued.

Back in Europe, however, plans were being implemented to adopt the modified Morris plan—now called the Maidenhead Grid Locator system. Officials in the three International Amateur Radio Union (IARU) regions were contacted about adopting the plan within their respective regions. Region 3 was the first to adopt the plan in 1982. Region 2 followed in 1983. Then finally, in April 1984, Region 1 adopted the Maidenhead Grid Locator system, with an implementation date of January 1, 1985.

With interest in the CSVHF Society awards program increasing, the American Radio Relay League (ARRL) formed an Ad Hoc committee to study the adoption of a League-sponsored awards program as a possible replacement for the CSVHF Society awards. During 1982, the committee, working closely with members of the board of the CSVHF Society, devel-

oped the VHF/UHF Century Club (VUCC) award, which incorporated the 100-grid concept. Seeing that the future lay in the Maidenhead locator system, the committee designed the program around it.

In January 1983, an article in *QST* by then Communications Manager John Lindholm, W1XX, announced the implementation of the awards program. Although the rules weren't spelled out entirely in the article (the rules for the higher microwave frequency awards were still being developed), a starting date of January 1, 1983 was set. The awards provided for decreasing requirements on increasingly higher frequencies. On both 6 and 2 meters the number of required grid locators to be worked was 100. On 125 and 70 cm, 50 grids were required. Operators on 902 and 1296 MHz needed 25 grids each. As the rules developed, requirements were spelled out for the higher bands. On 2.3 GHz the requirement was 10 grids. On the 3.4 GHz and above amateur bands the requirement was 5 grids. The recipients would receive *half century* awards for contacts on 125 (now 135) and 70 cm. For all bands above 70 cm, the recipients would receive *quarter century* awards. Initially, no award was offered for repeater or satellite contacts. However, on September 1, 1992, an award was created for working 100 grids via satellites. Endorsements were made available for working more grids on a particular band. However, no other endorsements (such as for mode or propagation path) have been authorized. (For more complete rules for the VUCC award and other VHF awards see Chapter 16.)

Interestingly enough, while this awards program

was designed to supersede the CSVHF Society program, it has done so only in *popularity*, as the CSVHF Society awards still exist today. However, according to Kent Britain, WA5VJB, one of the board members, no one has applied for any of the awards in the past ten years, and it's doubtful the mechanics would be in place to issue any if someone were to do so.

The VUCC program was an instant success, and the race was on to see who would be first to attain the award on each band. Keeping in step with the change, the June 1985 VHF QSO Party rules switched from sections to grid locators as multipliers, creating a whole new scoring procedure. Leaders were still based on sections and divisions. However, point accumulation was based on the grid locator, with its relatively uniform size, rather than the arbitrary size of different sections. Suddenly, a grid locator had the potential of becoming more rare than Delaware. Amateurs made efforts to be "in demand" by operating from one of these rare grid locators.

Over the years, because of the fleeting nature of VHF+ contacts, the grid locator has come to replace both the QTH and signal report information as the exchange for many VHF+ QSOs (the exception being meteor scatter and EME contacts, which still use a modified signal report system). However, this has not eroded what is considered a legitimate QSO. VHF+ operators have maintained a high sense of ethics. An operator agrees that he must hear both callsigns and the grid locator (or signal report on meteor scatter and EME QSOs), along with an acknowledgement from the other operator that he has also received the same information, in order for the QSO to be considered complete.

It's well accepted that if one operator hasn't received all the information necessary, the contact is incomplete and both operators will wait for another time to repeat the attempt. The "exchange a signal report only" type of QSO so prevalent on HF operations, and particularly on some DX nets, is rarely found on VHF.

Determining Your Grid Locator

The grid locator designator has up to six places—two letters followed by two numbers, followed by two letters. The starting grid field, the 20° by 10° large unit, begins with "AA." It's located between 180° and 160° W and 90° and 80° S. The next field—"AB"—is also found between 180° and 160° W, but is located between 80° and 70° S. Each successive field in that

column is ten degrees farther north, and is designated by an increasing second letter; the most northerly field is designated "AR." The next field to the east repeats the alphabetical sequence with the designator "BA," and continues the same progression north. The final field, found at the opposite corner (between 160° and 180° E and 80° and 90° N), is designated "RR."

Determining your grid locator location within the grid field is simply a matter of finding your longitude and latitude. There are one hundred 2° by 1° grid locators in each grid field. (Actually, these locations are more commonly referred to as "squares," as a throw back to the old CSVHF Society 1° by 1° system which, because of their equal degree dimensions, got mislabeled as squares. Unfortunately, the name "square" stuck.) Grid locators are numbered between "00" and "99." The lowest number is located at the

This array of microwave antennas tells why Greg McIntyre, AA5C, has VUCC on all bands through 10 GHz.

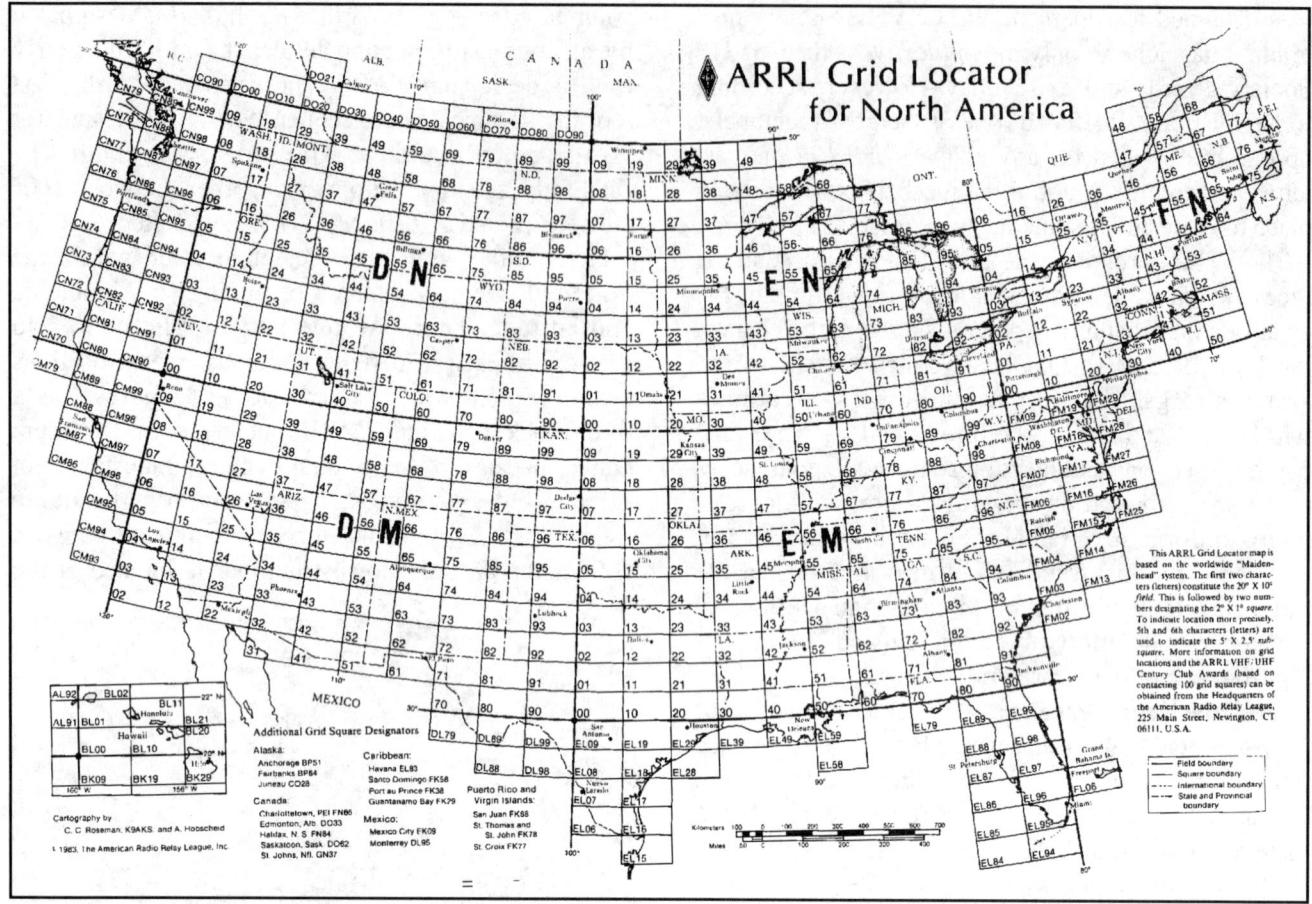

The ARRL Grid Locator map is based on the Maidenhead system and may be used as a handy reference. It is printed here courtesy the ARRL and may be ordered from them (see the Appendix).

southwest corner of the grid field and the highest number is located at the northeast corner. Find your two-digit number by counting how many degrees you are located east and north of the southwest corner. For example, if you're located within 100° and 80° W and 40° and 50° N, you are within grid field EN. If you're located between 96° and 94° W and 43° and 44° N, count one locator for each two degrees east of the corner and one locator for each one degree north of the corner. In this example, you'd be two locators east and three locators north of the corner—within grid locator EN23, or somewhere in northern Iowa or southern Minnesota.

Things get a bit more complicated if you want to figure out your exact designator. You must know your location down to the last 2.5 minutes. If your starting point is an even degree, increase the first letter by one increment for each 5 minutes of your degree intersection, beginning at the letter "A." If your starting point is an odd degree, increase the first letter by one increment for each 5 minutes east of your degree intersection, beginning at the letter "M." For each 2.5' north

of your degree intersection, increase the second letter by one increment. In neither the longitude nor the latitude designations will you find a letter beyond "X." Again in the above example, if the location fell between 95° 20' and 95° 25' W, and between 43° 15' and 43° 17.5' N, your grid locator would be EN23QG, and you would be located very near, if not at, the four-corner intersections of Osceola, Dickenson, O'Brian, and Clay Counties in Iowa. The last two letters are rarely used in North America, except to indicate much more accurately the locations in order to measure distance records (and then principally on the microwave bands). It's important to note that the system of finding your fifth and sixth designators is reversed when you are south of the equator and/or east of Greenwich longitude.

Those who don't want to bother figuring out their location and looking it up in an atlas can purchase the *ARRL Grid Square Map for North America* and the *ARRL World Grid Square Atlas* from the ARRL. The *Atlas* includes documentation for a BASIC program you can use to calculate your grid locator. Both items can be purchased directly from the League and may also be available at your local ham store.

If you have a IBM-PC type computer, there are shareware software packages available for downloading from various BBSs that will help you calculate the grid locator or the longitude and latitude if you know the other. They include BD, HAMGRID, LOCCALC, and SQUARES.

Determining Your Correct Longitude and Latitude

Finding out where you are on the Earth can be a bit of a challenge. There are two ways—the easy (but considerably more expensive) way and the slightly more difficult (but considerably cheaper) way. The easy way is to find someone with a Global Positioning System, or GPS, receiver. With one of these devices you can determine your longitude and latitude with a fair degree of accuracy.

These receivers rely on signals transmitted from at least three of the 24 satellites that are in synchronous orbit approximately 11,000 miles above the Earth. The GPS receiver must receive a minimum of three of these satellites to obtain the necessary signal to produce a two-dimensional location, which will allow the circuitry in the unit to calculate your location. However, it takes a minimum of four satellites for accurate elevation information. Some receivers are able to pick up signals from as many as six or more satellites.

There are several handheld models on the market from around $500 and up. One of them, the Trimble Navigation Scout, has the maidenhead grid locator system built into its software. Acquire this model, and your problem of finding out where you are in terms of your grid locator is solved.

You can use the U.S. Geological Survey 7.5 minute series topographic maps to plot your grid locator, but this is a slightly more difficult method. These maps are called 7.5 minute series because they cover an area 7.5' by 7.5'. They are available from the U.S. Geological Survey Centers in Denver, Colorado 80225 or Reston, Virginia 22092, or from the Oklahoma Geological Survey Center, Norman, Oklahoma 73069 or your local map store. I purchase mine from the local map store for $3 per map. Incidentally, the local map store is probably your best bet because each map is named and if you don't know the name of the map, you can at least look through their selection until you find the one you want.

Once you've acquired the right map, simply look for the correct location and read the scale on the side of the map to determine your longitude and latitude. For my QTH, the map is called "Britton, Oklahoma." I looked up my QTH and determined that I was approximately 1,200 feet west of the 97° 32' 30" mark and 5,200 feet north of the 35° 30' mark. The 97° × 35° indication places me within grid locator EM15. By a bit of interpolation and drawing on the map, I could see that I was within the 5' × 2.5' location bounded by 97° 30' to 97° 35' by 35° 30' to 35° 32' 30", which places me within the sublocator of "fm." Therefore, my grid locator is EM15fm.

Incidentally, once I knew my longitude and latitude, I could have plugged them into a software program and had it calculate my grid locator for me.

Why Are Grid Locators So Popular?

The quest for completing contacts with stations in different grids rivals that of county hunting on HF. To economize on space and time, anyone who wishes to go on a grid expedition will probably also want to put on as many grids as possible. Pat, W5OZI, Nick, W5FUA, and John, KB5IUA, held their 1992 grid expedition at the exact intersection of DM70, DM80, DL79, and DL89. The 1992 Chip Angle, N6CA, grid expedition took place at the exact location of the four corners of DM05, DM06, CM95, and CM96. The ARRL allows this sort of expedition, as long as the exact intersection can be determined by physical means. In both of these cases, the operators had the benefit of a survey marker.

Looking Ahead

What is the future of the grid locator system? As an incentive for activity, it still presents a challenge. There are 32,400 grid locators in the world. To date, Fred Fish, W5FF, has worked the most grids, close to 800. Fred has also worked all of the grids in two fields. Even so, his total accumulation represents only 2.3 percent of the total locators available. So even for the leader, there's still plenty of challenge left. When the VHF+ operators, who fled the bands for 10 meters because of the "easy" DX, return again, there will be even more active stations to work from different grid locators.

Chapter 16

VHF+ Contesting and Awards

Contesting on the VHF+ frequencies and on the HF frequencies is both alike and different. VHF+ contesting is like HF contesting in that the goal setting and preparation involved are the same. In addition, some software written for HF contests is also available for VHF+ contests. The principal difference between VHF+ and HF contesting is the length of time between contacts. For instance, when the band isn't open (on 6 meters) or when there's simply no one to work (on the higher VHF+ frequencies), you may wait for hours between contacts. On the other hand, you may begin a 90 QSO hour in the next 15 minutes.

Patience is one skill you must have on VHF+ that's not as necessary on HF. Knowing when to wait and when to quit is also very important.

VHF+ Contest Strategy

What does it take to create a winning contest station? To give you a picture of what it takes to win, I'll bor-row a bit from HF operators such as my friend John Dorr, K1AR, and a lot from the big guns of VHF+ contesting such as Dave Hallidy, KD5RO. Most of the advice that follows applies to the general VHF+ contests. However, some of the strategy also applies to Field Day and, to a lesser extent, the SMIRK contest.

Your Body

How you feel is just as important for the success of your contest station as the equipment it contains. Have you rested adequately before the contest? Even though VHF+ contests almost always allow you to get a good night's sleep (because the bands shut down at night), you still need to be in top shape for the endurance associated with contesting.

What are you eating? Some operators prefer a diet of pasta because it's high in carbohydrates. Taboos include caffeine (that includes chocolate) and sugar. (Some operators even avoid fruit because of the high

In a CQ WW VHF WPX Contest Mexico was represented by the special prefix XA5T (ops Rosendo, N5YBG, Enri-qaue, XE2FU, Mario, KF5RM). (Photo courtesy KF5RM)

Bernardo Gonzalez, XE2HWB, is shown here ready to make VHF+ contacts during a recent XF1G DXpedition. (Photo courtesy N6XQ)

Jack Henry, N6XQ, operated as XE2/N6XQ from this Vizcaino Peninsula, Baja California Sur, QTH while breaking the North American 10 GHz band DX record last year. (Photo courtesy N6XQ)

fructose content.) Both caffeine and sugar are stimulants that, after they wear off, could leave you in an energy crash. Don't try any new, especially spicy, foods just before the contest. Your digestive tract may not approve.

Your Goals

Your biggest decision is probably to define what it means to win. Winning for you may involve being top in the country for your category, or it may mean making ten contacts on 10 GHz. Setting a goal and writing it down helps you focus on what you intend to accomplish during the contest. Always remember: Whatever your goal, if you achieve it, you're a winner.

Your Station

What may seem to be the most obvious ingredient is also the most taken for granted. If you're contemplating operating a contest, your station *must* be in top shape. If you purchased a piece of new equipment recently, use it as much as it takes to become very familiar with it. Know the knobs and buttons, and their functions. Know the equipment's strengths and weaknesses. If possible, make several dozen contacts with it.

Make sure every aspect of the station is working to your satisfaction. Check the antenna, the coax, the rotator, and the rotator lead-in. Check the radios, the power supplies, the preamps, the amplifiers, and the cables used to connect all these items.

Are you operating away from your home QTH? Assemble your antennas and towers or masts just as you plan to use them in the field. There are two important reasons to go through the exercise of setting up everything. The obvious one is to make sure the equipment is working properly. The second is to ensure that you have overcome the learning curve associated with station assembly. The third is to make sure you have all the tools and parts necessary for construction once you get in the field. This, hopefully, will make station assembly in the field much easier and faster.

Once assembled, make several contacts with your "portable station" set up in the backyard. Be sure to make duplicate sets of cables—and check to see that they are working to your satisfaction. Also, make sure your field location is viable. Take a mobile or portable radio to the site. Check for power line noise or other reception problems. Make sure the lay of the land is sufficient to allow enough room for every station.

Look around your station. Is the setup comfortable? When you're sitting in your ergonomically correct chair, is everything within easy reach?

Are you using software or paper logs? If you've chosen to use software, make sure the computer is working and you really know the software (don't try to learn it the night before the contest). If you're using paper logs, make sure you have enough log sheets, dupe sheets, scratch paper, and pencils (yes, pencils, because they erase to make changes more easily).

Tim Marek, NC7K, and Dave Eubanks, NR6E, operated from Pond Peak, Nevada during the 1993 June VHF QSO Party. (Photo courtesy NC7K)

This is the EME station used by Hal Perry, KC4YO, during a recent ARRL EME contest. (Photo courtesy KM4XW)

This "mobile" 1296 MHz EME array netted Hal Perry, KC4YO, five contacts in three countries during the second weekend of an EME contest. (Photo courtesy KM4XW)

On The Air

VHF+ contests have an entirely different pace than HF contests. This sometimes frustrates HF operators who try contesting on VHF. Unless the band is open, you won't get the steady runs that HF contest operators experience. Therefore, it's imperative to tune through all the bands that you have available. This isn't as hard as it seems, because the other operator is just as motivated as you are to make multiple contacts.

Stay off the calling frequencies. If you have a loud signal people will come to you. If not, you can go to the loud signal. It takes only one station on the calling frequency to ruin it for everyone. When on 6 meters, stay out of the DX window. Reserve it for contacting DX stations only! During the 1987 June VHF QSO Party, operators in England were hearing stations as far away as the southwestern portion of the United States, but couldn't be heard because of stateside QRM in the DX window.

Have Fun

This hobby is supposed to be fun, so naturally contesting should also be fun. Unless you make it enjoyable for yourself, you'll find the whole experience frustrating and unfulfilling.

How can you make it fun? One way is by setting goals. You may want to work enough stations to earn a pin in one of the ARRL contests. You may want to work stations in new grids only. You may want to run QRP and see how many stations you can work.

Another way to make it fun is to find a group of like-minded hams and make the contest an outing (similar to Field Day). There's a group of contesters who gather each year at Dave Olean, K1WHS's QTH for the September contest. While they never run up a big score, they always have fun. (Word has it, however, that during the 1993 contest, in spite of their effort to keep a low score, they succumbed to the temptation of the best aurora conditions in years and ran it up.)

Part of the annual tradition is to go out to breakfast at Ding-a-Ling Cafe on Sunday morning. Lauren Libby, KXØO, one of the regulars at the contest site, reports that going to the Ding-a-Ling Cafe is an experience one will never forget!

Popular VHF+ Contests

Contests of interest to the VHF+ operator occur several times a year. The first one each year is the January VHF Sweepstakes. It is sponsored by the ARRL and includes all VHF+ ham bands.

The next are the VHF Sprints. They occur in April and May. They are short contests, lasting only four hours, and with the exception of the microwave bands, cover only one band at a time.

The most popular contest is the June VHF QSO Party. It occurs in early June and covers all the bands. It is most popular because it is summertime, when people are on vacation and sporadic-E is present on 6 and sometimes 2 meters.

A couple of weeks later the SMIRK organization

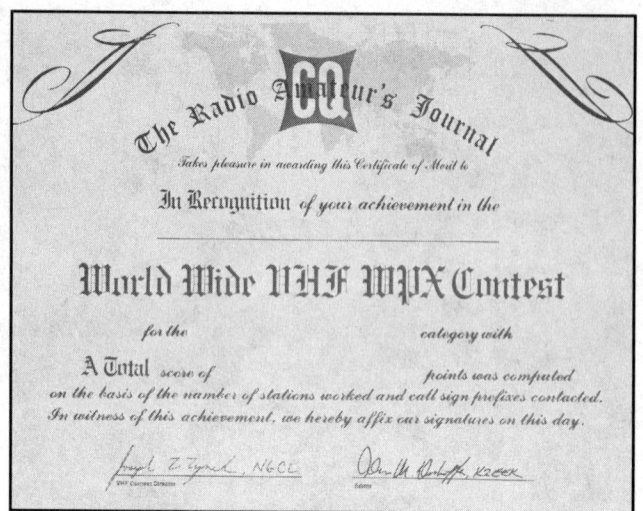

The CQ WW VHF WPX Contest is held in July each year and is sponsored by CQ magazine.

sponsors its annual 6 meter contest. For that contest you must exchange your SMIRK number as well as the grid locator. If you do not have a SMIRK number, simply tell the other operator you have none.

A week later is Field Day. This is the most popular contest in ham radio. It is important to the VHF+ operator because so many bonus points can be obtained by operating various specialties on the VHF+ bands. You can get bonus points for making packet contacts, for making satellite contacts, for having a dedicated VHF station and making contacts with it, and for having a dedicated Novice/Technician station and making contacts with it. So you see, there are many ways of exploring the VHF+ frequencies just during this one contest.

In July *CQ* has its annual VHF contest. It's a worldwide contest. However, it is undergoing changes, and the rules that governed the contest in the past may be changed for future contests.

In August there are a couple of specialized contests. The UHF contest occurs in the early part of the month. Then later in the month is the first weekend of the 10 GHz contest.

In September there are the Canadian Sprints, the ARRL VHF contest, and the second weekend of the 10 GHz contest. The Canadian Sprints are similar to the ARRL Sprints, and the ARRL VHF contest is similar to the June VHF QSO Party.

In October is the continuation of the Canadian Sprints and the first weekend of the ARRL EME contest. The first weekend of the EME contest is generally around the first weekend of the month, and the second weekend is sometimes the last weekend of

October or it can be the first weekend of November.

As you can see, there are plenty of contests in which you can participate. For summaries of the rules for each of these contests see the appropriate issues of *CQ* magazine.

How To Achieve VHF+ Awards

Completing the requirements for any amateur radio awards involves goal setting. The VHF+ frequencies are no different. Once you've made a decision to go to these bands, you must decide whether or not to pursue specific goals. If you do, there's a principal difference between HF and VHF—*time*. While, as you'll see below, some of the requirements for popular VHF+ awards can be met in as short a time as a weekend, some aren't met for decades.

Following are descriptions of available VHF+ awards, the most popular being the VHF/UHF Century Club (VUCC) award.

VHF/UHF Century Club (VUCC) Award

To recap information presented earlier in Chapter 15, the award requirements for VUCC are as follows: For 6 and 2 meters, you must work 100 grid locators; for 135 and 70 cm, the requirement is 50; for 33 and 23 cm, you must work 25 grids; for 13 cm, the requirement is 10; and for all other bands the requirement is 5.

Six meter operators from most parts of the country should have no difficulty obtaining VUCC. The regu-

The VHF/UHF Century Club (VUCC) award is based on the Maidenhead locator system and is available for working a specified number of locators on each band. It is sponsored by the ARRL.

lar existence of sporadic-E allows for fairly easy completion of contacts in distant grid locators. In fact, contest stations and other operators often contact stations in 100 different grid locators in one contest weekend.

The task is a bit more difficult on 2 meters. With a little luck, if you live in the center of a rather highly populated area and are also surrounded by highly populated areas, you should reach your goal pretty quickly. Here again, contest stations, and others, have been known to work stations in the required 100 grid locators in one weekend. However, if you live outside, or even on the fringe, of these high population areas, the task becomes exponentially more difficult. For example, if you live in the west, you're probably surrounded by vast expanses of land containing few hams, let alone VHF+ operators. Under these conditions, you must use extended forms of propagation to meet your goal. Meteor showers, sporadic-E, aurora, tropo, and EME become very important in these situations. Of all of these modes of propagation, meteor showers are probably the most predictable.

Those of you on a limited budget are probably least likely to choose EME as a way to meet your goal, but don't rule it out. As I mentioned in Chapter 14, San Hutson, K5YY, and Ray Soifer, W2RS, each presented excellent papers on EME made easy and relatively inexpensive at Central States VHF conferences. Both of these papers unlock some of the mystique of EME operation for the "little guy."

But I digress. We were talking about meteor showers. It's possible to make their predictability work for you. All you need to know is when the various meteor showers occur and what type of showers they are. Meteor showers are pretty well documented. In my column in *CQ* magazine, I list the current shower(s) for the month in the "VHF+ Calendar" sidebar. Additionally, many bulletin boards have software available that will help you predict when showers will occur.

An understanding of meteor shower types will unlock two more aspects of meteor shower propagation. These are how far you can work, and in what direction.

First, let's look at how far you can work. Remember, meteor burn-up ionizes the E-layer. This means that you're limited to distances ranging between 700 and 1,300 miles. Get a grid square map and draw a circle around your QTH that extends out to the 1,000 mile mark. Add two more circles at the 700 and 1,300 mile marks. These two additional circles represent the inner and outer limits for this mode.

In other words, within these two circles lie most of the possible grid locators you can work via meteor scatter. You'll notice, however, that there aren't 100 grid locators within these two circles. More on that later.

How do you make meteor scatter work for you? A quick glance at Chapter 11 will show you that the *Lyrids* peaks around April 21–22. Let's use it as an example. For this shower, the average height of ionization is around 65 miles. This means that the probability of completing a meteor contact drops off significantly at distances over 1,200 miles, and that contacts between 800 and 1,000 miles enjoy a high probability of completion.

What about direction? The *Lyrids* is a good north-south shower. Therefore, plan to work stations to your north or south.

Now you know two facets of operation, so find stations located within your circle and within the directions encompassing a high probability of completion. Tim Marek, NC7K, has produced a booklet entitled *North American VHF/UHF Directory*, which contains grid locators and phone numbers (where available) of many of the active weak signal VHF+ operators in North America. Contact Tim via his *Callbook* addresses for more information. Armed with this list, you can use it to contact hams and set up skeds.

How can you contact the remaining grid locators? The next most reliable mode of propagation is tropo. On any day, you can make contacts of between 100 and 150 miles with a modest station containing a 150 watt brick and a 15- to 17-element long boom beam between 25 and 30 feet in the air. You can increase your distance out to about 350 miles, or so, by what some call "brute force." However, it's necessary that both your station and the distant station are optimized for low noise and efficient power transfer to the antenna. It also helps if the other station is running high power. This is important not only so you can locate the station, but also so you know the precise direction in which to aim your antenna and return a signal back to the station in order to complete the contact.

Additional tropo contacts can be made in excess of 350 miles when the conditions are "just right." These conditions exist when a weather front is strategically located and causes the air to stabilize for several hours to several days, "trapping" the signals in a tropo "zone" over land or "duct" over water. Under the right conditions, distances nearing 1,000 miles can be reached over land, and in excess of 2,500 miles over water.

With the exception of the over-water path, however,

long-haul tropo only replicates the area already covered by meteor scatter. So, the principal area of concentration to increase the number of grid locators worked remains the close-in grids "skipped over" by meteor scatter.

Neither of the other forms of propagation—sporadic-E and aurora—are predictable with any kind of degree of reliability. Sporadic-E openings on 2 meters are most likely to occur during the months of May through July, with June being the peak. Some rare openings also occur during December and January. The only way to determine when sporadic-E is occurring is to observe the lower bands, such as 6 meters, commercial FM radio, or low VHF TV frequencies. It's then a matter of tracking the MUF (maximum usable frequency) until it reaches 2 meters and hoping that someone is on the air in the direction of the opening. However, sporadic-E, like long-haul tropo, only replicates the area already covered by meteor scatter.

As you saw in Chapter 12, aurora has its own limitations. It's a form of propagation that, when it occurs, tends to favor only the higher latitudes. On rare occasions it may track lower in the (northern) hemisphere. However, this may occur just once every couple of years. If you're trying to work your 100 grids within a limited time frame, especially if you live in the lower latitudes, don't even consider working via aurora. However, aurora can and should be considered a way of filling in some of those blanks on the grid locator map by those who live in the northern latitudes.

This discussion brings us back to the Moon. Although it takes more sophisticated equipment to get "on the Moon," the terrestrial distance is limited only to your common lunar window with the station you wish to work. Some who are close to their goal of working 100 grids have resorted to EME to get the last few. And, according to Ray Sofier, W2RS, there are plenty of high power stations with good receivers who can complete a contact with your station.

How about VUCC on 135 cm? For the past several years, this band has been neglected. Uncertainty has kept people off "in droves." Now that we know we have a protected segment, 135 cm should experience a resurgence in popularity.

The lower requirement of 50 grid locators for 135 cm reflects the lack of activity on the band, not the lack of propagation. Insofar as propagation is concerned, this band enjoys the best of both worlds: it shares propagation traits with its neighbors, 2 meters and 70 cm. Meteor scatter, albeit harder to work, does appear regularly on this band. EME exists, with slightly better conditions than 2 meters. Sporadic-E has been documented on very rare occasions. Tropo is considered by some to be better on this band, and aurora occasionally makes an appearance. The variety of propagation modes provides all the ingredients you need to work 50 grid locators.

Owing to its higher popularity, 70 cm presents a unique opportunity to achieve VUCC. Because more stations are on the air, some operators find that the 50 grid square requirement is actually easier to attain than the 100 grid square requirement for 2 meters—despite the reduced propagation opportunities on this band.

Let's look at what's available. First, meteor scatter does exist—although it takes considerably more patience to complete a contact. Unfortunately, many of the so-called "lesser showers" just don't produce the propagation on this band that they do on 2 meters. Also, owing to the dynamics of meteor scatter propagation, contacts in excess of 1,000 miles are very rare, indeed!

On 70 cm, tropo is the most popular way to fill in the blanks on the grid square map. As for 2 meters, tropo conditions exist regularly out to 150 miles. Brute force tropo can extend that to 350 to 400 miles. And when the band opens, tropo conditions can extend to more than double that range. Because of the nature of enhancement, tropo conditions often appear on 70 cm ahead of the lower bands.

EME can also be used to work additional grid locators. However, it's unnecessary to use this mode, because at least 50 grid locators fall within tropo limits.

As we go higher in the spectrum, it becomes more difficult to complete VUCC for the respective bands. The requirements for 33 and 23 cm are the same—25 grid locators.

Tropo is the chief form of propagation on both of these bands. There are more than enough grid locators within the tropo limitations. However, owing to the lack of population on the 33 cm band, ability to garner the necessary grid locators is much more difficult. It becomes necessary to work a station on another band and bring it with you to 33 cm. This is sometimes accomplished by working the station first on 23 cm!

Getting VUCC on all of the other higher bands involves soliciting help from your friends. Tropo is the chief form of propagation on these bands, and enough grids lie within average tropo limits for you to reach your goal. Unfortunately, not enough hams operate regularly from these grids on the required bands. Consequently, the *only* way to complete the requirements is to get others to activate the grids that you need.

The VUCC awards require a fee of $1 for the initial

application for each band, or for the satellite award. Subsequent endorsements require just an SASE. If you are a United States or Canadian applicant, you must have your QSL cards field checked by an official from an ARRL Special Service Affiliated Club, or the Radio Amateurs of Canada equivalent. To obtain the complete rules, application, information on grid locators, and a list of field checkers near you, send a letter and an SASE to the American Radio Relay League in the U.S. ((225 Main Street, Newington, Connecticut 06111) or to the Radio Amateurs of Canada (P.O. Box 356, Kingston, Ontario, K7L 4W2 Canada). Foreign applicants must have their QSL cards field checked by a designated official in their country.

Worked All States Award

To earn Worked All States (WAS) you must make contacts in all 50 states. The ARRL offers separate awards to those who work all states on 6 meters, 2 meters, 135 cm, and 70 cm, as well as for amateur satellite operation. WAS is open to ARRL members in the United States and Canada, and to other amateurs worldwide.

On 6 meters, depending on where you are located in the country and where we are in the solar cycle, it may take you between one summer and ten years to garner all the states. The most difficult to snag are Alaska and Hawaii. Those who live in the southeast will most likely have to wait until an F_2 propagation mode peak to work Alaska. However, it is possible to work Alaska *and* Hawaii on multi-hop sporadic-E. Nevertheless, these events are rare and you must watch for them.

On 2 meters you *must* rely on the Moon. Only if you live within a circle that includes the eastern halves of Nebraska, Kansas, and Oklahoma and the western halves of Missouri and Arkansas can you work all of the "lower 48." Even so, you still have to rely on EME for your Hawaiian and Alaskan contacts.

For the other two bands that have recipients of WAS awards, all of them have had to work several states via the Moon.

Cards and application forms can be verified locally. Check with your local radio club or the Awards Branch at ARRL Headquarters for the Awards Manager nearest you. Complete rules and application forms are available from the League for an SASE.

Worked All Continents (WAC) Award

To earn the Worked All Continents award, you must work six continents: Africa, Asia, Europe, North America, Oceania, and South America. This award is

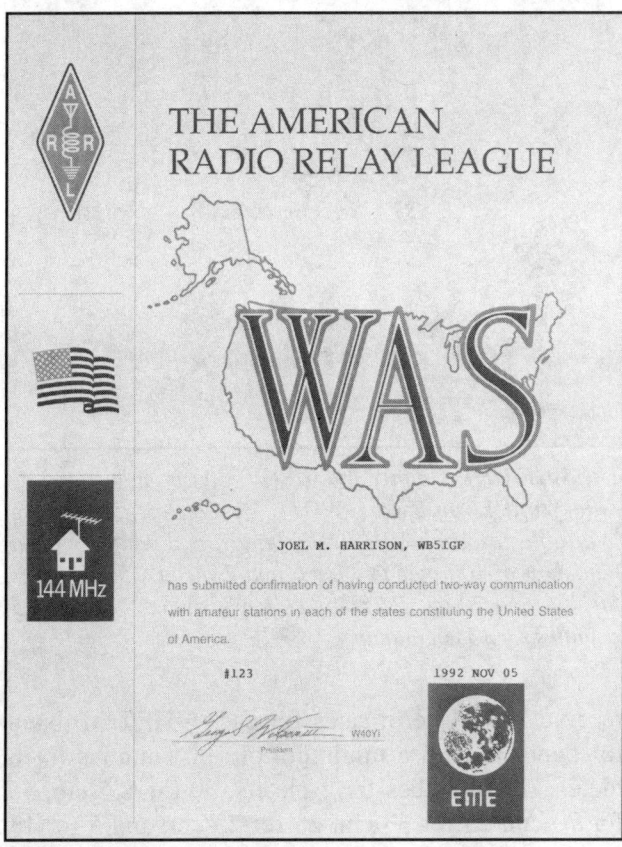

To earn the Worked All States award you must make contacts in all 50 states. The ARRL offers separate awards to those who work all states on 6 meters, 2 meters, 135 cm, 70 cm, and via amateur satellites.

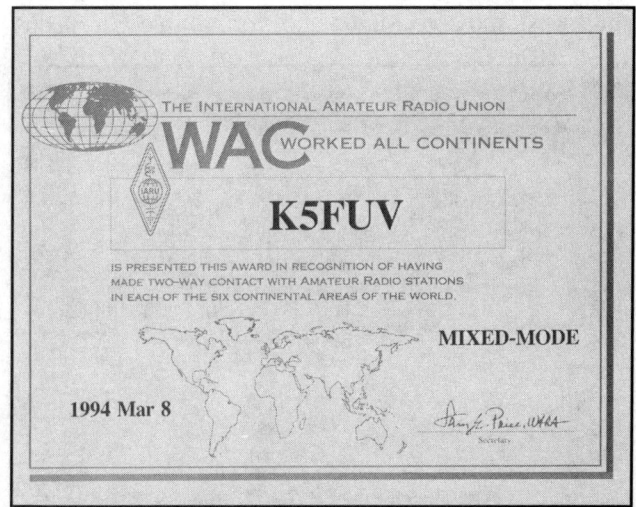

To earn the Worked All Continents award, you must work the six continents. A separate award is available for working all continents via amateur satellites. It is sponsored by the International Amateur Radio Union (IARU) and administered by the ARRL.

open to everyone, and is sponsored by the International Amateur Radio Union (IARU) and administered by the ARRL. The basic WAC award is offered with endorsements for 6 meters, 2 meters, and 70 cm, as well as for any higher band. A separate award is available for working all continents via amateur satellites.

In order to work WAC on any of the VHF+ ham bands you must consider EME contacts. While it is possible on 6 meters without the Moon, you will have to wait for a peak in the sunspot cycle for the F_2 propagation necessary for the long-haul contacts.

Complete rules and applications forms can be obtained from ARRL Headquarters for an SASE.

DX Century Club Award

To earn the DX Century Club (DXCC) award, you must work 100 countries. VHF+ DXCC awards have been issued for 6 and 2 meters, and satellite operation. It is open to ARRL members in the U.S. and Canada, and to other amateurs worldwide.

DXCC is not an award for the casual operator, even though this sunspot cycle has seen the first issuance of a number of DXCC awards for 6 meters. If you are *very* fortunate, you can work 50 countries via sporadic-E. The rest must be via some extended form of propagation, such as TE or F_2. A QTH in the right part of the country is a must, once again. Fortunately there's a DXCC rule that states you can count contacts made from anywhere in your home country; so if you like to travel, you can work all you can from the southwest, move to Maine and work all the Europeans

you can, then move to the west coast and work all the Asians you can. Theoretically, it's possible to achieve DXCC on the move! Here again, though, you must have quite a bit of luck. Nevertheless, if this is your plan, it will still probably take you at least ten years to reach your goal.

Two meters is the only other band on which anyone has accomplished DXCC. Only a handful of operators have achieved this milestone—and then only after investing thousands of dollars in their station and many years on the air.

For complete rules and application forms send an SASE to the ARRL.

VHF/UHF County Award

The VHF/UHF County Award (VUCA), sponsored by Side Winders on Two (SWOT), encourages VHF/UHF activity and recognizes operating achievements on the frequencies above 6 meters. You can

The VHF/UHF County Award (VUCA) is sponsored by Side Winders on Two (SWOT). VUCA certificates are available for each VHF/UHF band for contacts with a minimum of 100 counties for 6 and 2 meters, 50 counties for 135 and 70 cm, 25 counties for 33 and 23 cm, and 5 counties for 13 cm and up.

receive VUCA certificates for each VHF/UHF band for contacts with a minimum of 100 counties for 6 meters, 100 counties for 2 meters, 50 for 135 cm, 50 for 70 cm, 25 for 33 cm, 25 for 23 cm, and 5 for 13 cm and up. After earning the basic award, you can work toward endorsements for additional counties. VUCA is open to all licensed hams. Contacts made after January 1, 1983 count toward your total.

DX Century Club (DXCC) awards have been issued for 6 meters, 2 meters, and satellite operation. The award is sponsored by the ARRL.

Over 3,000 counties and independent cities in the United States qualify for this award. These counties are the same as those used for *CQ* magazine's USA-CA Counties Award. You can keep track of counties worked with *The CQ USA-CA County Award Record Book*, available from *CQ* for $2 each. The official county list is also published in *The ARRL Operating Manual*.

VUCA awards are numbered separately for each band. Endorsements are available for CW, SSB, EME, satellite, ATV, packet, FM simplex, mobile, and QRP operation. Contacts via terrestrial repeaters don't count.

Once you've made enough contacts to apply for VUCA, make sure they're recorded accurately in your record book. Your QSL cards must show the county or other identifiable location (city, town, latitude/longitude, etc.) for positive county determination. Ask two licensed amateurs to check your cards and sign the certification form found in the *Record Book* to verify that you have qualified. Your record book becomes your official SWOT VUCA record and won't be returned.

U.S. stations must make contacts from one county, or any county immediately adjoining it. Mobiles operating on a county line may be credited for no more than two counties at a time.

VUCA award fees are $4 for SWOT members and overseas stations, and $5 for U.S. non-SWOT members, for each certificate. Certificate endorsements are $1 and an SASE.

Send applications and make checks payable to L. G. Parsons, W5AL, SWOT VUCA Awards Manager, 3316 Edenburg Drive, Amarillo, Texas 79106, (806) 352-0835.

Other SWOT Awards

You can earn awards for working SWOT members. Work 10 members for the initial award. Successive awards are available for those who work members in increments of 25 (25, 50, 75, 100, and so on). Send $1 and SASE per award to Jerome Doerrie, K5IS, SWOT Membership Awards Manager, Rt. 2 Box 72, Booker, Texas 79005, (806) 658-2264.

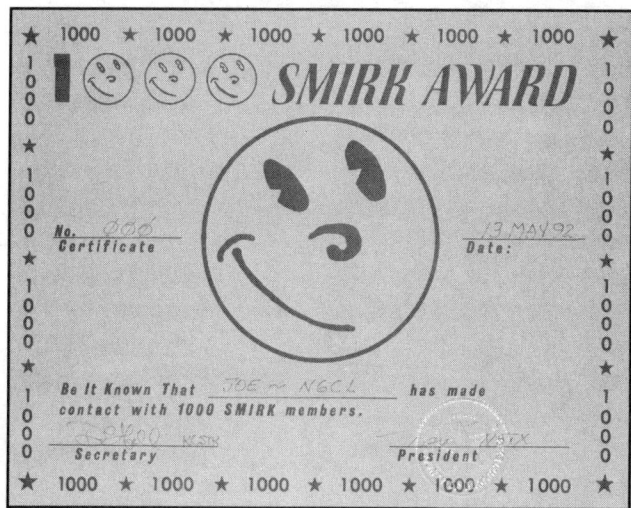

The Six Meter International Radio Klub (SMIRK) offers awards for working DX stations on 6 meters. Available are the DX Decade Club Award, the 50 Country Club Award, the 100 Country Club Award, and awards for working SMIRK members.

SMIRK Awards

The Six Meter International Radio Klub (SMIRK) offers awards for working DX stations on 6 meters. Contact 10 DXCC countries, and you can qualify for the DX Decade Club (DXDC) Award. Earn endorsements for each additional 10 countries you work. The 50 Country Club Award and the 100 Country Club Award are available for contacting 50 and 100 countries, respectively.

SMIRK also offers awards for working club members. You can earn the basic award, 100 SMIRK, for working 100 members. Advanced awards are offered for working 250, 500, and 1000 members.

All awards, with the exception of the 50 and 100 Country Awards, are self-certified. The 50 and 100 Country Club Awards require photocopies of your QSL cards.

Fees are as follows: $3 for 1000 SMIRK and DXDC, $5 for 50 and 100 Country Club. Seals for 100, 250, and 500 are $1 each.

To apply for these awards, contact Don Abell, KC5TK, SMIRK Awards Manager, 6821 West Avenue, San Antonio, Texas 78213, (512) 349-7234.

Chapter 17

Mountaintop and Rover Operation

Mountaintopping has long been popular on the VHF+ ham bands, mainly because the VHF+ operator recognized that the higher above obstructions he was, the farther he could transmit. In recent years, the grid locator system and the Rover category in contests have given this activity an added boost.

By definition, the Rover is a one- or two-person team that sets up portable operations from at least two grids during a set time period. Rovers have been known to operate from as many as 15 grid locators during a contest, but the average is closer to four or five grids on a given trip.

Rover contest operation has been around in one form or another for many years. Before grid locators, mobile stations would travel to rare states and put them on the air for others who needed to work them. Interest increased with the adoption of grid locators, as more mobile stations were able to travel to relatively nearby rare grid locators. (For example, from my home in Oklahoma City, I need only travel 45 miles, to the other side of El Reno, Oklahoma, to be inside EM05, a grid locator rarer than my home grid locator, EM15. However, the closest "rare" state is South Dakota, three states to the north.)

The Rover concept really got a push when contest station operators discovered that it provided a way to augment their operations during contests. Club stations such as the Rochester VHF Group, the Pack Rats, or the W2SZ Contest Group would enlist a mobile station to go to a nearby state, set up, and provide contacts to the club station.

The concept grew until, based on a recommendation from the ARRL Contest Advisory Committee, the League adopted a Rover category for the June 1991 VHF QSO Party. (The idea of including a Rover cate-

gory, as well as the limited multi-op category, was originated and promoted by Emil Pocock, W3EP, Curt Roseman, K9AKS, and Mike Owen, W9IP, in several of the "VHF-UHF Contesting!" columns in the *National Contest Journal*, beginning with the September/October 1989 issue. The idea really gained steam when the results of a survey they published in the March/April 1990 issue indicated the proposal had overwhelming support.) The category proved to be an instant success; nearly 50 stations submitted entries.

Riding their success, the League decided to adopt the category for the January and September VHF contests, as well. Based on feedback from fellow contest operators, I adopted the category for the revised CQ WW VHF WPX Contest. In the 1992 ARRL June VHF QSO Party contest, Rovers accounted for nearly 10 percent of the entries.

Clearly, there's a strong interest in this category. Interestingly enough, one of the reasons for creating it was to wean the captive Rover stations away from the super contest stations and encourage them to compete among themselves. Now the pendulum has swung to the other extreme. During the 1993 January VHF Sweepstakes contest, a group from the Hampden County Radio Association set up teams principally to work each other. In so doing, each team scored a previously unheard of 1.2+ million points. As of this writing, the League is considering revising the rules governing Rover operations. If you are interested in operating this category during a contest, check the current rules to make sure you're in compliance.

Although the Rover concept has become an official part of contests, the original activity—mountaintopping—is still popular. Most frequently during the summer months, a group of operators will get together and travel to rare grids to give others an opportunity to fill in the holes on their grid locator maps.

Why Is Mountaintopping/Rovering So Popular?

As I said before, it's relatively easy for most people to drive to a rare grid locator and put it on the air. And, with the inclusion of the category in contest rules, Rovers find themselves competing with each other to be the best. The fascination with Rovering also includes what I call the "being DX" factor. Once on the air, the Rover finds that contacts with his station are "in demand." Although not anywhere near as intense as an HF DX pileup, the thrill of being the "hunted" is just as real. Dave Hallidy, KD5RO, reports that one of his biggest thrills as a Rover is giving out new grids on 13 cm to operators who are new to that band and running minimal stations. Also, Rover operation is often relatively inexpensive and sometimes surprising. You never know what interesting events you may observe when you're out and about. For instance, Tim Marek, NC7K, reported that while he was setting up his equipment in a shopping center parking lot on a recent trip, he watched, open mouthed, as the local police busted a drug dealer right in front of him!

The Successful Rover Expedition

How do you put together a successful Rover expedition? To find out, I interviewed a number of success-

Dave Bostedor, N8NQS, is shown here working Europe from EN97.

ful operators and researched the operations of others. These intrepid Rovers included Ted Goldthorpe, WA4VCC, Gary Colborne, WA1EHL, Denise Hagedorn, AJØE, Tom Bishop, KØTLM, Dave Hallidy, KD5RO, John Walker, WZ8D, Emil Pocock, W3EP, Jerry Becker, WA8R, Byron Swainey, WA8NJR, Ron Hammil, KC6WLC, Tim Marek, NC7K, Kent Britain, WA5VJB, Frank Moorhus, AA2DR, Ray Veldran, N4KWX, John Lindholm, W1XX, Chip Angle, N6CA, Tom Brown, N7AMA, Pete Scola, WA7JTM, Pat Rose, W5OZI, John Godwin, KB5IUA, Wayne Overbeck, N6NB, Geoff Krauss, WA2GFP, and Bob Thompson, N4YZJ. A compilation of their experiences follows.

Perhaps the most important factor in all successful Rover trips is *planning*. This includes how far in advance you plan for your trip, where you go, how long you stay within a particular grid locator, what kind of equipment you take, what kind of vehicle (or vehicles) you drive, what time of year you take your trip, how long your trip is, and who goes with you. It helps to make a list of all of these items and check them off as you accomplish them.

First, consider your destination. To a large extent, where you go will be dictated by the rarity of the surrounding grids. For example, in the panhandle and western part of Texas there are several grids that are relatively rare. A Rover might start in DM96 and work his way down to DM91, spending most of the time in the most rare grid locator, DM94.

Unless you're out for a Sunday afternoon drive and you just happen to have the 6 meter rig in the car, you'll be better off if you have a good idea as to the location of the high points in the grids you plan to visit. This means you almost have to travel the route before your trip, or at least get in touch with someone who knows the area and has pictures. For instance, if you're looking for a place to go in Nevada, contact Tim Marek, NC7K, 360 Prestige Ct., Reno, Nevada 89506, or call him at (702) 972-4722. Tim has maps and photos of potential sites throughout the state.

Second, you must secure permission to operate from the sites you've selected. This means getting permission from property owners, local authorities, and so on, which may be very difficult. For example, Gary, WA1EHL, and Bob, N4YZJ, were run off their location on Rich Mountain during the 1992 June ARRL contest by local authorities because they were blocking an access road. Even though no one lived on property accessed by the road, a neighbor who didn't want them there complained. The authorities used the "access road blockage" as the basis for dispersing the Rover team. As disconcerting as that was, it doesn't compare with being hauled in for questioning for Roving around one's own neighborhood, as was Geoff Krauss, WA2GFP, during the January 1993 VHF Sweepstakes contest. Fortunately, his lawyering skills helped "bail him out" once he was at the police station.

After your trip, be sure to thank your host for allowing you to use the property for your hobby. It was no accident that Bryan, WA8MZQ, asked me to convey his appreciation to his hosts—including the Royal Canadian Air Force—in my column writeup of his 1992 trip.

How long should you stay at a given location? There are several factors to consider. Are you on an extended trip, or are you participating in a contest? Is the grid you're within relatively rare and worth the effort involved in spending some extra time? Are you in the contest to win, or are you just having a good time? If you are contesting, how long will you stay to work every last station before moving on? Is it worth it to you to stay just so someone can finally complete that contact on 23 cm—even though you're delayed an hour? During contests, some of the on-the-air intimidation from the intense operators on the other end can be relentless. Remember, you are in control. You make the decision when to pack up and move on. Denise, AJØE, and Tom, KØTLM, have often been faced with the decision to help someone complete a contact when time has expired for their schedule. On one occasion they stayed "too long" in an attempt to complete a contact (unsuccessfully). When they finally got on the road, they were only able to travel a short distance before being too exhausted to go any farther. They stopped at a motel and got rooms for the night. Unfortunately, they had used their credit cards to guarantee rooms at another motel farther down the road. Though they didn't lose a lot of money in the deal, it was nevertheless a bit disconcerting to have to pay to stay in two different places the same night.

What kind of equipment you take depends on how many bands you want to operate. For ease of operation, many operators choose out-of-the-box multiple-band radios. Those transceivers that offer the ability to operate on more than one band with a flick of the switch are very popular. Unfortunately, this benefit is also a drawback. If there's more than one person on your team, operating on one band will keep one of you very busy, while the other operators stand around awaiting a turn at the mike (or key). Plus, you may miss a band opening on one that you are not currently listening to.

If you plan to operate more than one band on the air simultaneously, take the necessary equipment.

You'll also need backup equipment. If something fails, you can't just drive home and replace it. Make sure you have extra radios, microphones, extra coaxial cable, and plenty of extra connectors. In your box of spare connectors include mike connectors, coaxial connectors, and phone plugs. Bring all the tools you think you'll need, and then some. What about a soldering iron? And don't forget the solder!

Antennas and rotators are a challenge. Some operators choose to mount the antennas atop their van. This saves set-up and disassembly time. However, this means the vehicle is the rotator. Others bring along a tower. It takes longer to put up the antennas, but they're also higher in the air. Still others opt for something in between, assembling the antennas on a single mast and attaching the mast to the van. They use an "armstrong" rotator (you know, your strong arm), or an inexpensive TV antenna rotator powered by the generator or an invertor (that's equipped to power a motor). John Lindholm, W1XX, has gone so far as to punch a hole in the top of his van, install a PVC type fitting, and run the mast down inside the van. When operating, he merely reaches over and "rotates" the antennas from the comfort of the van's interior.

The type of power you run tends to dictate how you'll run your equipment. If you run more than a brick (100 to 150 watts), you'll need either a small gas generator or deep discharge marine type batteries. Notice, I said batteries. Even if you use just a brick, a marine battery is something to consider. It was awfully mortifying to find myself standing by the road holding my jumper cables in the air on Sunday morning during the 1992 June contest. Fortunately, I didn't have to answer too many embarrassing questions once a very kind motorist stopped to assist me. Faced with similar experiences in the past, others have decided to keep the engine running in their van all the time.

How do you keep track of your contacts? You can use a lap top computer. However, most of the people I've talked with who have tried this method have returned to pencil and paper logs. The chief complaint is that the software used isn't versatile enough to accommodate the logging needs of the Rover, and the computer is just one more item that can break. If you go with paper logs, bring plenty of pencils and paper. Have a safe place to stow the logs. Keep containers full of sharpened pencils near the operating positions, because you never know when you'll drop one or break a lead.

Shown here is Randy Simons, NØLRJ's FM operating setup with his "Portaple" antenna up only 18 feet or so.

How do you keep track of time? If you're going to run meteor schedules, you must have an accurate source of time for the sequencing. There is available a clock which can lock onto the WWV signals on 10 MHz, but if that luxury isn't in your budget, then you'll need an HF radio and an antenna that will pick up the WWV signal, plus a clock on which you can control the "seconds" setting.

What's the best way to operate CW? The cheapest way is a hand key. However, some operators use keyers that can double as beacons. The choice is up to you. Remember to bring a hand key as a backup. You never know.

What kind of vehicle makes a good Rover station? The most popular seems to be a full-size passenger or panel van. Chip, N6CA, and Wayne, N6NB, both have elaborately equipped panel vans for this kind of operation. Denise, AJØE, and Tom, KØTLM, use a passenger van that has been partially converted for camping out. Gary, WA1EHL, and Bob, N4YZJ, use a small travel trailer. Ted, WA4VCC, and Itice, KB4CSE, use a passenger van, and set the equipment between the two captain's chairs. Gordon West, WB6NOA, and Carmine Fiorello, AB6KE, chose to use yet another kind of vehicle during the 1992 June VHF QSO Party. They operated in the Pacific aboard Gordon's yacht!

Whatever the vehicle, it's imperative that it be in excellent running condition (down to the tires) for the trip. Your automobile club towing service will never find some of the locations you choose. Also, know how to operate what you are driving. Towing something takes a certain set of skills. Driving something with limited side and rear vision takes another set of skills. If you're driving something with a limited field of vision and towing something else, your work is really cut out for you.

In addition to choosing the type of vehicle you plan to use, you must decide where you're going to stay. This refers back to planning your route. If you stay inside the vehicle, be sure you're protected from the elements. If you choose to stay at a motel, know where to find one once you arrive at your destination.

What about operating while in motion? If you're in a contest, the temptation exists to make as many points as possible. One source of points is FM simplex (if you are near a metropolitan area that supports this type of activity). The other is via 6 meters, if the band is open. I operated while traveling from grid locator to grid locator, using the FT-726, a brick, and the base mast of a Hustler whip mobile antenna from my car, and the same setup with a 2 meter whip on the van. However, it's safer to have a co-pilot do the operating. Trying to operate contest style and log at the same time can be very distracting, not to mention dangerous. If you choose to operate and drive, and find yourself in a pileup, pull over. If you don't, you might find yourself in another type of pileup.

What time of year is the best time for Roving? Obviously, the best time is when the band is open. This

NØLRJ's "Super Rover" is in the traveling position and is almost finished. All beams are in rotatable mounts for vertical or horizontal orientation. The 2 meter and 220 MHz beams elevate for satellite, meteor, and EME operation. The top section snaps off for other configurations, and the entire array lifts out for field installation.

often seems to be during the summer. However, when planning for your trip, make provisions for any kind of weather. You may run into snow in June, if you travel through some parts of North America.

Whom will you take on your trip? You may want to go by yourself. However, the long stretches between band openings make for lonely times. If you choose to travel with someone, make sure you are compatible before you go on a long trip. Not only is compatibility important, so is trust. I watched a video tape that showed Jerry, WA8R, and Byron, WA8NJR, walking in front of the pickup that John, WZ8D, was driving very slowly across a rickety old bridge in the middle of nowhere in northern Canada. Trust is knowing that your friends are going to lead you across the right spot. Trust is also knowing that your friend isn't going to run over you.

How many of you should there be on a trip? If

On a dirt road somewhere in southwest Oklahoma two crazy hams—Jim Rudnicki, NZ7T, and your author—set up a Rover station in a broken-down 1975 Chevy panel van and, among other things, "stole" a farmer's TV signals (see text). Mountaintopping and Rovering are among the most fun activities on the VHF+ ham bands.

you're operating in a contest, all contest rules (as they are written now) state that there may be no more than two operators. However, for the *CQ* contests and (with clarification from Billy Lunt, KR1R, at the League) the ARRL contests, a third person, a non-operator, can go along as a driver. Obviously, if you're not participating in a contest, take along as many of your friends as you can live with at a time!

During the planning stages of your trip, check out your equipment exactly as it is to be used. When he operated from VP5, Chris, WA2HMK, discovered the need for this in a dramatic way. Once at VP5, he found things didn't go together quite as he expected. On your trip, you'll find the same thing. Knowing what to expect before you leave can save you from yet another headache when you arrive at your destination.

Probably the most important part of your planning is giving someone your itinerary. Without it, your friends won't know where to look if something happens to you. Your family, not to mention your many friends on the VHF+ frequencies, care about you. They want you to have fun on your trip. But, they also want you to return home—or at least know how to find you if you don't make it back.

What about publicity? When you have your trip itinerary worked out, let me know and I'll publicize it in my column. Remember, the more publicity, the more successful your trip.

The following story, written by my friend Jim Rudnicki, NZ7T, first appeared in the "The Dummy Load," his local newsletter in Utah. It is used here with the permission of Jim to illustrate just how much fun Rovering can be, even for the novice!

Rovering From Cement

Before you dismiss this article as another tadpole/dipole fish story, let me explain. The FAA in its infinite wisdom had sent me off to Oklahoma City for two weeks of fun in the sun. Not wanting to sit around on the weekend with nothing to do, I decided to learn about the vagaries of sporadic-E radio wave propagation on the VHF bands.

I had the pleasure of meeting Joe Lynch, N6CL, the VHF editor of *CQ* magazine. Joe was kind enough to invite me along as a guest operator for his Rover operation during the 1993 ARRL June VHF QSO Party.

A Rover is a mobile station with one or two operators who operate on as many VHF/UHF bands as possible from as many Maidenhead Grid Locations as possible during a VHF Contest. While this sounds well and good, my experience from this weekend defines Roving as two crazy hams careening all over central Oklahoma in an old van, operating while mobile, searching for a quiet out-of-the-way spot, setting up antennas, contesting, packing up, then going off over hill and dale to repeat the process several times!

The day started innocently enough; Joe picked me up in his van around 10:00 a.m., and we proceeded southwest into central Oklahoma. At this point there didn't seem to be any real urgency to this Roving at all. We stopped after a while to put oil, gas, and water into the van.

Joe asked, "What time is it?"

After I told him it was 20 minutes before contest time, the pace accelerated dramatically. Joe started quickly hooking up the Yaesu FT-726, amplifiers, and SQLOOP and whip antennas.

Innocently I asked, "Are you going to operate while you drive?"

Joe replied, "Of course not, you are!"

"I am?" I asked. "I don't even know anything about this type of contest!"

After the initial shock wore off, we were on our way making contacts. Joe was navigating in and around Chickasha looking for a quiet spot that wouldn't upset the local farmers, and I was leaning sideways out of my seat operating the FT-726, which was on the floor between us. Which brings me to the first major question of the day: How in the heck do you hold the VFO knob still while driving on a washboard road, much less hear what's going on on the radio?

After our first few contacts on 6 meters, I was well into the swing of things. Six meters was open via sporadic-E propagation into the east coast, and we were making contacts with just a mobile whip. With luck like that, I couldn't wait until we stopped and put up the beam! But our luck didn't hold. Near Verden (grid locator EM05) we stopped and put up the beam, but by that time the band had gone dead. Really dead. That's why it's known as "sporadic-E"!

Oh, well. Not to worry. We just packed everything back up and headed back into Chickasha for more gas, oil, water (for the van), and some burgers (for us)! Contesters do get hungry!

Our next stop was near Cement, OK in grid locator EM04. Cement is a *very* small town, and we didn't have to try very hard to find a quiet spot. We set up shop on a quiet dirt road between some wheat fields, lashed our 2 meter beam to a large convenient fencepost, and put the 6 meter beam on a tripod on top of the van.

This time, after the beams were up, 6 meters was wide open. We made well over a hundred contacts on 6 meters, and also made several contacts on 2 meters. It was truly amazing how good the propagation was on 6 meters. We worked stations from New York to Nevada, and as far north as Ontario, Canada. The only problem we had was that after all that driving around I had completely lost my sense of direction. (Those of you who have known me for a long time probably are not surprised!)

At one point Joe asked me to point the 2 meter beam towards Dallas.

"OK, fine." I said. "Which way is Dallas?"

I had no clue. Later in the evening when the stars appeared, I was able to find my way around.

During our stint in Cement, the owner of the aforementioned convenient fencepost showed up. What he found were two crazy hams, one hollering into a radio "CQ CQ Contest N6CL N6CL EM04 EM04!" and another ham hanging out the door twisting the antennas on the roof.

The farmer was very nice and polite as he said, "I know it's none of my business, but what in the heck are you doin'?"

We kindly explained in nice general terms what we were up to. The farmer said he had never heard of anything like that, but he just wanted to be sure we weren't "stealing his TV signals"!

"Oh, yes, and by the way, that's my gate you're using to hold up your antennas," the farmer said.

"Hmm," I said. "Would you like us to move?"

"Nahh, y'all just have your fun," said the farmer.

With a rather confused look in his eye, off he drove, and we got right back to work!

Work we did! Six meters yielded dozens of multipliers well into the late-night hours. Joe operated and I logged, and visa versa.

Around 11 PM we decided we'd had enough for one day, and we took down the antennas. Taking antennas down when you're all tired out is one thing. Taking antennas down in the pitch dark when you're *very* tired is another! Because we were out in the middle of nowhere, we just placed the antennas out in the middle of the road. Together we dismantled the masts, coiled coax, and threw everything in the back of the van.

After a very long and tiring day, I arrived back at my apartment at 12:30 a.m. ready to fall down and sleep for a week.

Joe, on the other hand, ventured out again on Sunday, activating two more grids, and even made some 70 cm contacts with a 2 meter mag mount that he placed on the side of the van for horizontal polarization!

Over all, it was a very successful contest. We made 389 contacts in 224 grids, operating from grids EM04, 05, 16, 06, and 15. A short breakdown follows: 6 meters, 295 QSOs, 171 grids; 2 meters, 84 QSOs, 46 grids; and 70 cm, 10 QSOs, 7 grids.

The moral of this story is that there are many facets of amateur radio. Don't be afraid to find about the other modes and bands, and find out what folks are up to! I was lucky enough to ride with an expert *(Hah!—N6CL)* in this field, and pack a lot of learning into a long day. Of course now I have caught the bug for VHF operating, and my budget may never be the same!

One of the last things Joe told me was "I hope to hear you on 2 meters in August during the *Perseids* meteor shower." Meteor shower? Hmm, I guess that will be another story. . . . 73 es happy operating, Jim Rudnicki, NZ7T.

Chapter 18

The Art of QSLing

As a 33 year amateur radio veteran, I've watched the evolution of QSLing. When I was first licensed as a kid, I wanted to exchange QSL cards with every ham I worked. You could even QSL eyeball QSOs. (In those days there was an organization that sponsored awards for so many eyeball QSOs, WAS eyeball QSOs, and even DXCC eyeball QSOs, but it died a natural death a few years after it started.)

After repeaters became the vogue and logging requirements were eliminated, I went through a bit of withdrawal as I learned not to QSL everyone I worked (especially if I worked them through a repeater). I guess we all felt that a repeater QSO was somewhat meaningless and not worth the effort of a card—particularly if we worked the same station day after day.

In some sense, it appears that this attitude has almost put an end to the practice of QSLing in general.

However, there are exceptions. Operators on 160 meters, the WARC bands, and VHF+ frequencies seem to be more apt to QSL or, at least, to answer one.

I've found that approximately 70-plus percent of the operators to whom I send cards respond with one of their own. The percentage increases when I include an SASE (a self-addressed stamped envelope; more on this later). It seems that VHF+ operators in general are still interested in QSLing.

Why bother QSLing, anyway? Not withstanding that often printed phrase seen on many QSL cards "The final courtesy of a QSO is a QSL," is it still worth the effort? Judging by the mail I receive, yes.

How do you QSL successfully? There seem to be some unwritten rules adhered to more or less by those who respond to cards they receive. Let's look at these practices to discover how to successfully receive cards. Keep in mind that these rules will only increase your

One of the author's QSL cards pictures him "all dressed up for our recent QSO." QSL cards aren't expensive, so there's no reason why you can't have a decent card. Be sure to include your grid locator information on your card, too.

I got all dressed up for our recent QSO.

N6CL

73 From
Joe Lynch
P.O. Box 73
Oklahoma City,
OK 73101 USA

chances, not guarantee a response. There are some operators who refuse to keep a log, let alone QSL anyone they work. You'll soon find out who these people are when you attempt to get a card from them.

It's important to have a decent QSL card of your own. I have two. One pictures me in a tuxedo with the caption "I got all dressed up for our recent QSO"; the other, used by those connected with *CQ* magazine, identifies me as the VHF editor. QSL cards aren't expensive, so there's no reason why you can't have a decent card. Make sure, however, that the QSO information is on the same side of the card as your callsign. If not, the recipient will have to flip your card back and forth to look up the data. Also, be sure to include grid locator information on your QSL card. It's quite a chore to track down a grid locator using an atlas and a magnifying glass.

Fill out your QSL card as soon as possible after the contact. If you're lazy (like me) and wait around, your memory of the contact (and maybe even the log book entry) may disappear. The longer you take to start the QSLing process, the less likely you are to receive a card in return.

Make sure to fill out your card properly. Use either a computer printed label or *legible* handwriting. The correct date and time are absolutely critical. There's some debate as to what order the date and month should appear. Most, however, prefer the universal order of date, month, year.

Write the date in standard (or Arabic) numerals, and the month in an abbreviated spelling of its name or in Roman numerals. Print the year out in full in Arabic numerals, or abbreviate it by listing the last two digits of the year. It's important to make a distinction between the date and the month. Using Arabic numerals for both the date and the month leads to confusion, particularly when the date and month are listed as "10/2." This creates doubt as to whether the QSO took place on February 10th or October 2nd.

There's also some justification for using Roman numerals for the month. The spelling or printing of the month may be confusing to someone else, particularly an operator from another country. However, because of the universality of Roman numerals, there's little doubt as to the character's meaning.

If your card doesn't already specify "2X" or something else to indicate that the mode was the same both ways, indicate this when you enter the mode, unless, of course, it wasn't the same both ways. It's not as critical on the VHF+ frequencies because most operators aren't concerned about the mode; they are just concerned that the QSO took place.

On HF, however, mode is important. Many awards offer endorsements for "all SSB" or "all CW." In some instances, cross-mode contacts are disallowed for award credit. Therefore, it becomes imperative to indicate how the contact was completed both ways. If both of you used the same mode during the contact, this information is usually indicated by writing "2X" or "2 way."

For some, correct indication of the signal report is most important. This is especially true if you work someone on meteor scatter and use "S-2," or work someone on EME and use "M" or "O" as the report. (It caused some consternation at the DXCC desk the first time those HFers saw the VHFers' signal reports when operators applied for their 2 meter DXCC!)

Don't forget to sign the card or use a signature stamp. The signature attests to the fact that the information is correct. The ARRL and other QSL card checkers treat these cards almost as though they are legal documents and look for that signature to prove authenticity.

Once you've filled out your card properly, mail it. If you let it sit around, you might misplace it, and you'll have to make out another card.

How should you mail your card? Should you stick a stamp on the card and mail it? Should you stick the card in an envelope and pay the additional postage? Should you include an SASE? It depends on whether or not you want the other operator's card. If you're responding to someone else's card, you can simply put a stamp on yours and mail it. However, be aware that your card may be eaten up or torn to shreds before it makes it to the other end—if it arrives at all. To protect your card, put it in an envelope.

Now that you have it inside the envelope, should you include an SASE? If you need the other operator's card, include one—even if the other operator needs yours! I've heard of two instances when both operators during a QSO acknowledged to each other on the air that they needed the other's grid locator and, thus, a QSL card. Assuming that the other operator would simply return a card out of courtesy, the first operator sent his without an SASE. After more than a year, the first operator ran into the second operator at a hamfest. When queried about receiving a card, the second operator acknowledged receiving the card, but added that he wasn't going to send a return card until he received an SASE! When in doubt, include an SASE.

If you do use an SASE, make sure your name and

address are properly and legibly printed on the return envelope and attach the proper postage. I once received a card, with an SASE, from an operator in Europe who needed confirmation from Oklahoma. He had hand printed his name and address on the return envelope. I used his envelope to send my card back to him. His envelope was eventually returned to me as undeliverable!

If you're mailing a card to someone in your own country, you know how much postage is required. However, if you're mailing a card to someone in another country, check with the post office or other sources—such as DX newsletters or magazines—to determine the current amount of postage required for that country.

How do you pay for the return postage? There are services that sell mint postage for particular countries. You can also use International Reply Coupons (IRCs) available from the post office. IRCs are also available for around 50 to 60 cents from individuals who advertise their availability. By international agreement, one IRC pays for postage from another country equivalent to one unit of airmail postage from your country. However, this doesn't always work in practice. Check with your local post office to be sure.

You can also pay for postage with currency. Often referred to as "green stamps," a U.S. dollar bill is sometimes offered as payment for the return trip. There are two problems with this method. First, it gets expensive. Second, mail to some countries is regularly subject to pilfering. Your dollar bill may be extracted and the remaining contents discarded long before it ever reaches its intended destination.

You can also QSL a DX station by sending your card to the other station's manager. This usually increases the likelihood of receiving a card in return. How do you obtain the name of someone's QSL manager? Listen to the station on the air or look up the name in "The Go List." Published by Jan, K6HHD, and Jay O'Brien, W6GO, this is the most comprehensive list in the world. You may subscribe by mailing $2.50 for one issue or $25 for a full year to The GO List, P.O. Box 700, Rio Linda, California 95673-0700 (check with them for the most current prices). If you're on packet, you may be able to access their data base, as many SYSOPs subscribe to "The GO List" data base service. "The VHF QSL List"—previously maintained by Harry Schools, KA3B, and now published by N6CL—is another source for locating QSL managers. For a copy of the most recent list, send an SASE with your request to Joe Lynch, N6CL, P.O.

Box 73, Oklahoma City, Oklahoma 73101.

You can also QSL via QSL bureaus. Most VHF+ QSLing is domestic and ineligible for bureau use. However, if you work DX and want to take advantage of the bureaus, here's how.

Most countries maintain bureaus for use by their radio amateurs. Addresses for these bureaus can be found in *The Radio Amateur Callbook*. Each country has its own regulations concerning eligibility for using its bureau. When in doubt, QSL direct.

In the United States, the ARRL maintains two separate bureaus—incoming and outgoing. The incoming bureau is run by volunteers within your call-letter location. The outgoing bureau is located at League headquarters.

The information in the following section on the ARRL QSL Bureau was furnished by the American Radio Relay League. It has previously appeared in various issues of *QST* and appears here courtesy of *QST* and the ARRL. Information on the Canadian QSL Bureaus appears here courtesy of the Radio Amateurs of Canada.

The ARRL Incoming QSL Service

Purpose. In the United States, the ARRL DX QSL Bureau System is made up of numerous call area bureaus that act as central clearing houses for QSLs arriving from foreign countries. These "incoming" bureaus are staffed by volunteers. The service is free and ARRL membership is not required.

How It Works. Most countries have "outgoing" QSL bureaus that operate in much the same manner as the ARRL Outgoing QSL Service. The member sends his cards to his outgoing bureau, where they are packaged and shipped to the appropriate countries. A majority of the DX QSLs are shipped directly to the individual incoming bureaus, where volunteers sort the incoming QSLs by the first letter of the callsign suffix. One individual may be assigned the responsibility of handling one or more letters of the alphabet. Operating costs are funded from ARRL membership dues.

Claiming Your QSLs. Send a 5 × 7 1/2 or 6 × 9 inch self-addressed, stamped envelope (SASE) to the bureau serving your callsign district. Neatly print your callsign in the upper left-hand corner of the envelope. It is suggested that you send envelopes with first-class stamps already affixed, and clip extra postage to each envelope. Then, if you receive more than 1 ounce of cards, they can be sent in a single package. Some incoming bureaus sell envelopes or postage credits in addition to their normal SASE handling duties. They provide the proper envelope and postage upon the pre-payment of a certain fee. To learn the exact arrangements for prepayment send your inquiry with a SASE to your area bureau. A list of bureaus appears later in this chapter.

Helpful Hints. Good cooperation between the DXer and the bureau is important to ensure a smooth flow of cards. Remember that the people who work in the area bureaus are volunteers. They are providing you with a valuable service. With that thought in mind, please pay close attention to the following dos and don'ts.

• DO keep self-addressed 5 × 7½ or 6 × 9 inch envelopes on file at your bureau, with your call in the upper left corner, and affix at least one unit of first-class postage.

• DO send the bureau enough postage to cover SASEs on file and enough to take care of possible postage rate increases.

• DO respond quickly to any bureau request for SASEs, stamps, or money. Unclaimed card backlogs are the bureau's biggest problem.

• DO notify the bureau of your new call as you upgrade. Please send SASEs with your new call, in addition to SASEs with your old call.

• DO include an SASE with any information request to the bureau.

• DO notify the bureau in writing if you don't want your cards.

• DON'T send domestic US-to-US cards to your call area bureau.

• DON'T expect DX cards to arrive for several months after the QSO. Overseas delivery is very slow. Many cards coming from overseas bureaus are over a year old.

• DON'T send your outgoing DX cards to your call area bureau.

• DON'T send SASEs to your "portable" bureau. For example, AA2Z/1 sends SASEs to the W2 bureau, not the W1 bureau.

• DON'T send SASEs to the ARRL Outgoing QSL Service.

ARRL Incoming DX QSL Bureau Addresses
First Call Area, all calls*: W1 QSL Bureau, Y.C.C.C., Box 216, Forest Park Station, Springfield, Massachusetts 01108.

Second Call Area, all calls*: ARRL 2nd District

QSL Bureau, N.J.D.X.A., P.O. Box 599, Morris Plains, New Jersey 07950.

Third Call Area, all calls: C-CARS, P.O. Box 448, New Kingstown, Pennsylvania 17072-0448.

Fourth Call Area, all single-letter prefixes (K4, N4, W4): Mecklenburg Amateur Radio Club, P.O. Box DX, Charlotte, North Carolina 28220.

Fourth Call Area, all two-letter prefixes (AA4, KB4, NC4, WD4, etc.): Sterling Park Amateur Radio Club, Call Box 599, Sterling Park, Virginia 22170.

Fifth Call Area, all calls*: ARRL W5 QSL Bureau, P.O. Box 50625, Midland, Texas 79710.

Sixth Call Area, all calls*: ARRL Sixth 6th District DX QSL Bureau, P.O. Box 1460, Sun Valley, California 91352.

Seventh Call Area, all calls: Willamette Valley DX Club, Inc., P.O. Box 555, Portland, Oregon 97207.

Eighth Call Area, all calls: 8th Area QSL Bureau, P.O. Box 182165, Columbus, Ohio 43218-2165.

Ninth Call Area, all calls*: Northern Illinois DX Assn., Box 519, Elmhurst, Illinois 60126 .

Zero Call Area, all calls*: WØQSL Bureau, P.O. Box 4798, Overland Park, Kansas 66204.

Puerto Rico, all calls*: KP4 QSL Bureau, P.O. Box 1061, San Juan, Puerto Rico 00902.

U.S. Virgin Islands, all calls: Virgin Islands ARC, GPO Box 11360, Charlotte, Amalie, Virgin Islands 00801.

Hawaiian Islands, all calls*: Wayne Jones, NH6GJ, P.O. Box 788, Wahiawa, Hawaii 96786.

Alaska, all calls*: Alaska QSL Bureau, 4304 Garfield St., Anchorage, Alaska 99503.

Guam, all calls: MARC, Box 445, Agana, Guam 96910.

SWL: Mike Witkowski, WDX9JFT, 4206 Nebel St., Stevens Point, Wisconsin 54481.

The Canadian Incoming QSL Service

QSL cards for Canada may be sent to the following addresses. Note that only members of Radio Amateurs of Canada are allowed to use these bureaus.

RAC DX QSL Bureau System, Kennebcasis Valley Amateur Radio Club, Box 51, St. John, NB E2L 3X1.

QSL cards may also be sent to the following individual bureaus.

VE1, VE9, VEØ, VY2*: L. J. Fader, VE1FQ, P.O. Box 663 Halifax, NS B3J 2T3.

These bureaus sell envelopes or postage credits. Send an SASE to the bureau for further information.

VO1, VO2: Roland Peddle, VO1BD, P.O. Box 6, St. John's, NF A1C 5H5.

VE2: A. G. Daemen, VE2IJ, 2960 Douglas Ave., Montreal, PQ H3R 2E3.

VE3: Garry Hammond, VE3XN, 5 McLaren Ave., Listowel, ON N4W 3K1.

VE4: Adam Romanchuck, VE4SN, 26 Morrison St., Winnipeg, MB R2B 3V4.

VE5*: B. J. Madsen, VE5FX, 739 Washington Dr., Weyburn, SK S4H 2S4.

VE6*: Norm Waltho, VE6VW, P.O. Box 1890, Morinville, AB T0G 1P0.

VE7*: Dennis Livesay, VE7DK, 8309 112th St., Delta, BC V4C 4W7.

VE8*: Rolf Ziemann, VE8RZ, 2 Taylor Road., Yellowknife, NWT X1A 2K9.

VY1: W. L. Champagne, VY1AU, P.O. Box 4597, Whitehorse, YU Y1A 2RB.

The ARRL Outgoing QSL Service

One of the greatest bargains of League membership is being able to use the ARRL Outgoing QSL Service to send your DX QSL cards overseas to foreign QSL Bureaus. *(Note: The ARRL QSL Service should not be used to exchange QSL cards within the 48 contiguous states.)* Your ticket for using this service is your *QST* address label and just $2 per pound of cards. For those not quite so DX active (sending 10 cards or less), enclose $1. You can't even get a deal like that at your local post office! The potential savings over the substantial cost of individual QSLing is equal to many times the price of your annual dues. Your cards are sorted promptly by the outgoing service staff, and cards are on their way overseas usually within a week of their arrival at ARRL Headquarters. Approximately two million cards are handled each year!

QSL cards are shipped to QSL Bureaus throughout the world. These bureaus are typically maintained by the national Amateur Radio Society of each country. While cards are not sent to individuals or individual QSL managers, keep in mind that what you might lose in speed is more than made up for in the convenience and savings of not having to address and mail QSL cards separately. (In the case of DXpeditions and/or active DX stations that use U.S. QSL managers, a better approach is to QSL directly to the QSL manager. The various DX newsletters, the "W6GO QSL Manager Directory," and other publications, are good sources of up-to-date QSL manager information.)

As postage costs become increasingly prohibitive, don't go broke before you're even halfway towards making DXCC. There's a better and cheaper way— "QSL VIA BURO" through the ARRL outgoing QSL Service.

How To Use The ARRL Outgoing QSL Service. (1) Presort your DX QSLs alphabetically by parent callsign prefix (AP, C6, CE, DL, F, G, JA, LU, PY, 5N, 9Y, and so on). When sorting countries that have multiple prefixes, keep each country's prefixes grouped together in your alphabetical stack. Addresses are not required. DO NOT separate the country prefix using paper clips, rubber bands, slips of paper, or envelopes.

(2) Enclose the address label from your current copy of *QST*. The label shows that you are a current ARRL member.

(3) Enclose payment of $2 per each pound of cards; there are approximately 150 cards to a one pound. A package of 10 cards or less costs $1. Pay by check (or money order) and write your callsign on the check. Send "green stamps" (cash) at your own risk.

(4) Include only the cards, address label, and check in the package. Wrap the package securely and address it to the ARRL Outgoing QSL Service, 225 Main Street, Newington, Connecticut 06111.

(5) Family members may also use the service by enclosing their QSLs with those of the primary member. Include the appropriate fee with each individual's cards and indicate "family membership" on the primary member's *QST* address label.

(6) Blind members who do not receive *QST* need only include the appropriate fee along with a note indicating the cards are from a blind member.

(7) ARRL affiliated-club stations may use the service when submitting club QSLs by indicating the club name. Club secretaries should check affiliation papers to ensure that the affiliation is current. In addition to sending club station QSLs through this service, affiliated clubs may also "pool" their members' individual QSL cards to effect an even greater savings. Each club member using this service must also be a League member. Cards should be sorted "en masse" by prefix, and a *QST* label enclosed for each ARRL member.

Countries Not Served By The Outgoing QSL Service. Approximately 260 DXCC countries are served by the ARRL Outgoing QSL Service, as detailed in the ARRL DXCC Countries List. This includes nearly every active country. As noted previously, cards are forwarded from the ARRL Outgoing

Service to a counterpart bureau in each of these countries. In some cases, there is no incoming bureau in a particular country and cards cannot be forwarded. However, QSL cards can be sent to a QSL manager—i.e., 3C1MB via (EA7KF). For this reason, the ARRL Outgoing Service cannot forward cards to the following countries: A5 Bhutan, A6 United Arab Emirates, A7 Qatar, C9 Mozambique, D2 Angola, EP Iran, ET Ethiopia, J5 Guinea-Bissau, KC4 U.S. bases in Antarctica, KC6 Belau, V6 (KC6) Micronesia, KH1 Baker and Howland Is., KH4 Midway Is., KH5 Palmyra and Jarvis Is., KH7 Kure Is., KH8 American Samoa, KH9 Wake Is., KH0 Mariana Is., KP1 Navassa Is., KP5 Desecheo Is., OD Lebanon, P5 North Korea, S2 Bangladesh, T2 Tuvalu, T3 Kiribati, T5 Somalia, TJ Cameroon, TL Central African Republic, TN Congo, TT Chad, TY Benin, TZ Mali, V4 (VP2K) St. Kitts & Nevis, VP2E Anguilla, VP2M Montserrat, VQ9 Chagos, VR6 Pitcairn Island, XT Burkina Faso, XU Kampuchea, XW Laos, XX9 Macao, 1Z (XZ) Myanmar (Burma), YA Afghanistan, ZA Albania, ZD7 St. Helena, ZD9 Tristan da Cunha, ZK3 Tokelau, 3C Equatorial Guinea, 3C0 Pagalu Is., 3V Tunisia, 3W, XV Vietnam, 3X Guinea, 5A Libya, 5H Tanzania, 5R Madagascar, 5T Mauritania, 5U Niger, 5X Uganda, 7O, 4W Yemen, 7Q Malawi, 8Q Maldives, 9G Ghana, 9N Nepal, 9Q Zaire, 9U Burundi.

Further Notes

SWL cards can be forwarded through the QSL Service. The ARRL no longer hold cards for countries without an incoming bureau. Only cards indicating a QSL manager for a station in these particular countries will be forwarded.

The Radio Amateurs of Canada also maintains an outgoing QSL bureau. It is again only open to members of RAC. Contact them for further information and rules. The address is: RAC National Outgoing QSL Bureau, Bag 5000, Morinville, Alberta T0G 1P0 Canada.

Recommended QSL Card Dimensions

The efficient operation of the worldwide system of QSL bureaus requires that cards be easy to handle and sort. Cards of unusual dimensions—either much larger or much smaller than normal—slow the work of the bureaus, most of which is done by unpaid volunteers. A review of the cards received by the ARRL Outgoing QSL Service indicates that most fall in the following range: height $2\,3/4$ to $4\,1/4$ inches (70 to 110 mm), width $4\,3/4$ to $6\,1/4$ inches (120 to 160 mm). Cards in this range can easily be sorted, stacked, and packaged. Cards outside this range create problems; in particular, the larger cards often cannot be handled without folding or otherwise damaging them. In the interest of efficient operation of the worldwide QSL bureau system, it is recommended that cards entering the system be limited to the range of dimensions given. (Note: IARU Region 2 has suggested the following dimensions as optimum: height $3\,1/2$ inches [90 mm], width $5\,1/2$ inches [140 mm].)

Have Patience!

Be patient. It takes time to process cards—particularly if you're waiting for a card from someone who receives plenty of them. When I returned from my trip to Montserrat a couple of years ago, I had over 200 pieces of mail waiting for me. Eventually, I received over 350 requests for cards out of 1,180 contacts. After ordering cards, entering all the data in my computer, sorting the incoming cards, printing the labels, and mailing everything, four months had passed. During that time I received two duplicate requests for cards.

Patience may mean waiting up to a year, or more, for a return card. But someday, when you least expect it, your mailbox will hold that long-awaited card.

Surprisingly, in spite of the fact that several stations no longer QSL, the topic is still very popular. There's something special about getting anything (other than a bill) in the mail. And it's especially enjoyable to receive mail from someone with whom you are now friends.

Chapter 19

And Finally . . .

Most of you who read my column in *CQ* magazine know that I like to end on a positive note, usually interlacing it with some positive advice. Well, this book is no different. Therefore, I present the following on integrity.

Your Goals and Integrity

In Chapter 16, I discussed goal setting as it relates to the VHF+ frequencies. Coupled with goals is integrity. Integrity is a very personal and individual trait. It may have been imprinted by religion or by a highly moral upbringing, but however it was imparted, it's a part of each one of us.

Most of us include honesty in our definition of integrity. Honesty, as it relates to goal setting, basically asks the question, "Did you play by the rules in order to achieve your goals?" If your goals include working toward completion of the requirements of the various awards, such as those outlined in Chapter 16, the rules for these awards are clearly defined and reflect the honesty you apply toward completion of these requirements.

Let's look at two aspects of these requirements: the definition of a QSO and the FCC requirement to "use the minimum power necessary to complete the contact."

First, what is a QSO? Our friends on HF have refined the QSO and, especially in contests, have streamlined the definition to mean that you hear another station, send your call and a signal report, and the other station does the same. In some net operations, it's become even more streamlined. The net control tells each station the other's call and asks simply that they exchange signal reports. The contact lasts mere seconds, and the net control declares that a contact is "complete" when he hears both operators

correctly repeat the other operator's signal report—even though it's sometimes clear that one operator has simply guessed at the signal report of the other.

We on VHF+ have slightly different standards. We will not accept a QSO as complete until both operators acknowledge to each other that they have received both a signal report, or a grid locator, or some other mutually agreed upon exchange of information, and the complete calls of both stations.

When did this different standard develop? Some believe that when what I call "fractional" (by "fractional" I mean the contact takes place over a predetermined, mutually agreed upon period of time, and by bits and pieces at a time) QSOs started taking place, a definition of what was considered to be a QSO had to be specified.

Probably one of the earliest examples of this was the first 2 meter meteor scatter contact, which took place between Paul Wilson, W4HHK, and Tommy Thompson, W2UK. As this mode of propagation was experimental, there was no definition of what was considered a QSO. Therefore, Paul and Tommy looked to the League, specifically to Ed Tilton, W1HDQ, then editor of *QST*'s "The World Above 50 Mc." column, to define what was necessary for a completed contact. Ed determined that both operators had to acknowledge to each other that they had received both calls and the correct signal report; the latter had to be confirmed by repeating the signal report received back to the other operator.

Reliance on Ed's definition led to the rejection of their first claimed contact in August 1953. It wasn't until the second contact that both Paul and Tommy received enough information from each other for Ed to consider the QSO complete.

Over the passage of time, the definition of a QSO

has undergone little change. The only minor modification is that the signal report received need not be repeated back to the other operator. A simple acknowledgement, using the word "Roger" on voice or "R" on CW, is considered sufficient.

Nowadays, we don't send information on every contact to the VHF editors of the various magazines or newsletters or to the sponsors of awards. We simply certify that we did make a complete contact or have complied with the rules. It's up to us to maintain our integrity and complete the requirements according to the rules and to play by the unwritten rules of what's considered to be a QSO.

The other aspect I want to look at in relation to integrity is the "minimum power" requirement. We've all heard the expression "California kilowatt." It has come to mean someone who runs in excess of the legal limit—particularly when chasing DX on the HF bands. Unfortunately, you would be fooling yourself to think that such "kilowatts" don't exist on VHF.

We need to ask ourselves, "On whatever band the operator used, what did he accomplish when running excess power?" If excess power is the only way to make the contact, then what did the operator really prove? The rest of the world still considers the achievement impossible under the "rules."

Ultimately, one's integrity dictates that either the rules are kept or they are not. If an amateur radio operator keeps to the rules, his accomplishments are real, he will be regarded as someone who can be emulated, and his goals will be considered feasible. If not, then doubt continues concerning the viability of the accomplishment.

I feel fortunate to observe a high degree of integrity within the VHF+ community. Rarely do I hear accounts of "skirting" the rules in order to complete the requirements. I am proud to know this, and I hope that such integrity will pervade all of ham radio.

In Closing

In preparing this book, I've learned a tremendous amount about the wonderful world of VHF+. I have also learned a great deal by attending VHF+ seminars and by your correspondence sent to me as the VHF editor for *CQ* magazine. I hope you have learned, too, by reading this book.

I also hope you've enjoyed reading this book as much as I enjoyed writing it for you. Certainly, the last three years as *CQ*'s VHF editor have been the most fun time of my ham radio career—especially because I've met so many of you in person after contacting you on the VHF+ frequencies. If I haven't had the pleasure of meeting you, I hope to do so very soon. For now, though, 73, Joe Lynch, N6CL.

Appendix

VHF+ Ham Bands, Calling Frequencies, and Nets

The world of VHF+ carries with it an enormous amount of frequency real estate. In addition, the complexity of HF band plans pales in comparison to the recommendations established on VHF+. What follows are VHF+ band summaries, calling frequencies, and HF nets used for VHF+ operation.

VHF+ Ham Bands

The VHF+ ham bands extend from 50 MHz to light. Following is a list of the ham bands, authorized modes, generally accepted band plans, and operating privileges for each of them.

All Technician class licensees, regardless of having passed the Morse code test, and all other license classes above Technician are authorized to use all privileges on all the VHF+ frequencies. Novice class licensees are authorized full access to the 222–225 MHz ham band and a portion of the 1240–1300 MHz ham band. It is expected that the Amateur Radio Service will be authorized packet operation within a new frequency allocation of 219–220 MHz in the near future, so this assumption is made in the list. Finally, it is anticipated that the frequencies of 449–450 MHz will be shared with wind profiler operations. (Wind profilers are radar instruments that show wind speed at various levels of the atmosphere, and thus a profile of the wind at a given moment.) Therefore, this assumption is made when listing the various portions of the 70 cm band.

For information about repeaters in your area, consult your local VHF/repeater coordinating organization or the *ARRL Repeater Directory*.

50–54 MHz

Operation on the 50–54 MHz band is open to all licensees holding a Technician class or above license.

50.000–50.100 MHz: This portion of the band is exclusively for CW operation. The section of the band between 50.000 and 50.020 MHz is unofficially reserved for EME operation. The section of the band between 50.060 and 50.080 MHz is reserved for unattended U.S. beacon stations, although foreign (non-U.S.) and attended U.S. beacons can be found anywhere within the CW portion of the band. Rarely do stations who use this portion of the band for CW contacts operate below 50.090 MHz.

50.100–50.300 MHz: This portion of the band is for SSB operation, although CW contacts are permitted and do sometimes take place. The section of the band between 50.100 and 50.125 MHz is reserved for DX contacts only. Domestic contacts within this window are greatly discouraged.

50.300–50.600 MHz: This portion of the band is open for all modes of operation.

50.600–50.800 MHz: This portion of the band is reserved for non-voice communications, like packet.

50.800–51.000 MHz: This portion of the band is reserved for radio control operations, for controlling model airplanes or cars. There is a 20 kHz separation between each frequency used for such operations.

51.000–51.100 MHz: This portion of the band is reserved for a Pacific and European DX window. Because some countries have not granted privileges below these frequencies, DXers know to look here for some DX opportunities.

51.120–53.980 MHz: With a few exceptions (listed below), this portion of the band is used for FM and repeater operations. The frequencies of 51.500–51.600, 52.020–52.040, 52.520–52.540, and 53.000–53.020 MHz are all reserved for FM simplex. The frequencies of 53.100, 53.200, 53.300, 53.400, 53.500,

53.600, 53.700, and 53.800 MHz are all reserved for radio remote-control operations.

144–148 MHz

Operation on this band is open to all licensees holding a Technician class or above license.

144.000–144.100 MHz: This portion of the band is exclusively for CW operation. The section of the band between 144.000 and 144.050 is reserved for EME operation. The section of the band between 144.050 and 144.100 is reserved for both EME and weak signal CW operation.

144.100–144.275 MHz: This portion of the band is for SSB operation, although CW contacts are permitted and do sometimes take place. The section of the band between 144.100 and 144.200 MHz is used for weak signal and EME SSB operation. The section of the band between 144.200 and 144.275 MHz is used for general SSB operation.

144.275–144.300 MHz: This portion of the band is for beacon stations.

144.300–144.500 MHz: This portion of the band is for amateur satellite operations.

144.500–144.600 MHz: This portion of the band is for linear translator inputs.

144.600–144.900 MHz: This portion of the band is for FM repeater inputs.

144.900–145.100 MHz: This portion of the band is for weak signal and FM simplex operation.

145.100–145.200 MHz: This portion of the band is for linear translator outputs.

145.200–145.500 MHz: This portion of the band is for FM repeater outputs.

145.500–145.800 MHz: This portion of the band is set aside for experimental activities, such as communication between the space shuttle and Earth.

145.800–146.000 MHz: This portion of the band is for satellite operations.

146.010–146.370 MHz: This portion of the band is for FM repeater inputs.

146.400–146.580 MHz: This portion of the band is for FM simplex operation.

146.610–147.390 MHz: This portion of the band is for FM repeater outputs.

147.420–147.570 MHz: This portion of the band is for FM simplex operation.

147.600–147.990 MHz: This portion of the band is for FM repeater inputs.

219–220 MHz and 222–225 MHz

Operation on the 219–220 MHz portion of this band is open to all licensees holding a Technician class or above. Operation on the 222–225 MHz portion of this band is open to all licensees, regardless of license class. However, certain power restrictions apply. Within the 219–220 MHz portion stations may not operate more than 50 watts output. Within the 222–225 MHz portion of the band Novice class licensees may not operate with more than 25 watts PEP output.

Notice: As of this writing, the FCC has not authorized use of the 219–220 MHz portion of the band. The information contained here reflects the most probable regulations when authorization is given. Please consult up-to-date information before operating on this portion of the band.

219.000–220.000 MHz: This portion of the band is reserved for high speed point-to-point packet operations.

222.000–222.150 MHz: This portion of the band is reserved exclusively for non-repeater type operation. By mutual nationwide agreement, only CW and SSB operations take place within this portion of the band.

222.150–222.250 MHz: This portion of the band is for general-purpose CW and SSB operation and, by local agreement, FM repeater inputs.

222.250–223.380 MHz: This portion of the band is for FM repeater inputs.

223.400–223.520 MHz: This portion of the band is for FM simplex operation.

223.530–223.640 MHz: This portion of the band is for digital and packet operations.

223.640–223.700 MHz: This portion of the band is for link and control operations.

223.710–223.850 MHz: This portion of the band is for simplex and packet operations and for FM repeater outputs.

223.850–224.980 MHz: This portion of the band is for FM repeater outputs.

420–450 MHz

Operation on this band is open to all licensees holding a Technician class license or above. However, certain geographic power restrictions apply to operations on this band. Consult the FCC rules to see if your location is affected by these restrictions.

420.000–426.000 MHz: This portion of the band is for ATV simplex or repeater operations, with 421.250 MHz being the video carrier frequency. This portion is also used for experimental and control links.

426.000–432.000 MHz: This portion of the band is for ATV simplex operation, with 427.250 MHz being the video carrier frequency.

432.000–432.070 MHz: This portion of the band is reserved exclusively for EME operation.

432.070–432.100 MHz: This portion of the band is reserved for CW weak signal operation.

432.100–432.300 MHz: This portion of the band is reserved for CW and SSB weak signal operation.

432.300–432.400 MHz: This portion of the band is reserved for beacons.

432.400–433.000 MHz: This portion of the band is reserved for CW and SSB weak signal operation.

433.000–435.000 MHz: This portion of the band is reserved for auxiliary and FM repeater links.

435.000–438.000 MHz: This portion of the band is set aside for satellite operations throughout the world.

438.000–444.000 MHz: This portion of the band is for ATV repeater inputs; 439.250 MHz is the video carrier frequency.

442.000–445.000 MHz: This portion of the band is reserved for repeater inputs and outputs, on a local option.

445.000–447.000 MHz: This portion of the band is shared by auxiliary and control links, repeaters, and simplex operation, on a local option.

447.000–449.000 MHz: This portion of the band is reserved for FM repeater inputs and outputs, on a local option.

449.000–450.000 MHz: This portion of the band is reserved for FM repeater inputs and outputs, on a local option. However, it may also be shared with wind profilers in your area. Consult your local coordinating organization before establishing transmitting equipment on this portion of the band.

902–928 MHz

Operation on this band is open to all those holding a Technician class license or above. However, certain geographic power restrictions apply. Consult the FCC rules to see if your location is affected by these restrictions. Pending FCC proposed rulemaking may affect major portions of this band. Due to the uncertainty of the band's future, most band plans and repeater operations have been suspended. Therefore, the listings below are ones which *recommend* activity.

902.000–904.000 MHz: This portion of the band is reserved for narrowband, weak signal operation.

902.000–902.800 MHz: This portion of the band is reserved for SSTV, FAX, ACSSB, and experimental operation.

902.300–902.400 MHz: This portion of the band is reserved for beacons.

902.800–903.000 MHz: This portion of the band is reserved for future EME and CW expansion.

903.000–903.050 MHz: This portion of the band is exclusively reserved for EME operation.

903.050–903.100 MHz: This portion of the band is reserved for CW operation.

903.100–903.400 MHz: This portion of the band is reserved for CW and SSB operation.

904.000–906.000 MHz: This portion of the band is reserved for digital transmissions.

906.000–909.000 MHz: This portion of the band is reserved for FM repeater outputs.

909.000–915.000 MHz: This portion of the band is reserved for ATV operation.

915.000–918.000 MHz: This portion of the band is reserved for digital transmissions.

918.000–921.000 MHz: This portion of the band is reserved for FM repeater inputs.

921.000–927.000 MHz: This portion of the band is reserved for ATV operation.

927.000–928.000 MHz: This portion of the band is reserved for FM simplex and link operations.

1240–1300 MHz

Operation on the 1240–1300 MHz portion of this band is open to all those holding a Technician class license or above. Operation on the 1270–1295 MHz portion of this band is open to all licensees, regardless of license class. However, within this portion, Novice class stations may not operate more than 5 watts PEP output. Certain geographic restrictions apply to operations on this band. Consult FCC rules to see if your location is affected by these restrictions.

1240.000–1246.000 MHz: This portion of the band is reserved for ATV operation.

1246.000–1248.000 MHz: This portion of the band is reserved for NBFM point-to-point and digital operations.

1248.000–1252.000 MHz: This portion of the band is reserved for digital communication operations.

1252.000–1258.000 MHz: This portion of the band is reserved for ATV operation.

1258.000–1260.000 MHz: This portion of the band is reserved for NBFM point-to-point and digital operations.

1260.000–1270.000 MHz: This portion of the band is reserved for satellite uplink operations. It's also reserved for wideband experimental and simplex ATV operation.

1270.000–1276.000 MHz: This portion of the band is reserved for FM repeater input operations.

1276.000–1282.000 MHz: This portion of the band

is reserved for ATV operation.

1282.000–1288.000 MHz: This portion of the band is reserved for FM repeater output operations.

1288.000–1294.000 MHz: This portion of the band is reserved for wideband experimental and simplex ATV operation.

1294.000–1295.000 MHz: This portion of the band is reserved for narrowband FM simplex operation.

1295.000–1297.000 MHz: This portion of the band is reserved for narrow bandwidth weak signal operation (no FM).

1295.000–1295.800 MHz: This portion of the band is reserved for SSTV, FAX, ACSSB, and experimental operation.

1296.000–1296.050 MHz: This portion of the band is exclusively reserved for EME operation.

1296.050–1296.100 MHz: This portion of the band is reserved for CW operation.

1296.070–1296.080 MHz: This portion of the band is reserved for beacons.

1296.100–1296.300 MHz: This portion of the band is reserved for CW and SSB operation.

1296.300–1296.400 MHz: This portion of the band is reserved for beacons.

1296.400–1296.800 MHz: This portion of the band is reserved for cross-band linear translator operation.

1296.800–1297.000 MHz: This portion of the band is exclusively reserved for experimental beacons.

1297.000–1300.000 MHz: This portion of the band is reserved for digital communication operations.

2300–2310 MHz and 2390–2450 MHz

Operation on this band is open to all those holding a Technician class license or above. Weak signal operations are centered around ±100 kHz of 2304.100 MHz.

Notice: Pending proposals by the National Telecommunications and Information Agency may strip away major portions of this band from amateur use. Consult up-to-date information and regulations before operating on this band.

2300.000–2303.000 MHz: This portion of the band is reserved for high speed digital communications.

2303.000–2303.500 MHz: This portion of the band is reserved for packet operation.

2303.500–2303.800 MHz: This portion of the band is reserved for packet operation.

2303.800–2303.900 MHz: This portion of the band is reserved for packet, TTY, CW, and EME communications.

2303.900–2304.100 MHz: This portion of the band

is reserved for CW and EME operations.

2304.100–2304.200 MHz: The portion of the band is reserved for CW and SSB operations.

2304.200–2304.300 MHz: This portion of the band is reserved for SSB, SSTV, ACSSB, FAX, packet AM, and AMTOR communications.

2304.300–2304.320 MHz: This portion of the band is reserved for propagation beacon networks.

2304.320–2304.400 MHz: This portion of the band is reserved for propagation beacons.

2304.400–2304.500 MHz: This portion of the band is reserved for SSB, SSTV, ACSSB, FAX, packet AM, and AMTOR experimental.

2304.500–2304.700 MHz: This portion of the band is reserved for crossband linear translator inputs.

2304.700–2304.900 MHz: This portion of the band is reserved for crossband linear translator outputs.

2304.900–2305.000 MHz: The portion is reserved for experimental beacons.

2305.000–2306.000 MHz: This portion is reserved for FM simplex (25 kHz spacing).

2306.000–2309.000 MHz: This portion of the band is reserved for FM repeater inputs (25 kHz spacing).

2309.000–2310.000 MHz: This portion of the band is reserved for fast scan TV operations.

2396.000–2399.000 MHz: This portion of the band is reserved for control and auxiliary links.

2400.000–2410.000 MHz: This portion of the band is set aside for satellite operations around the world, with 2403.000–2408.000 MHz reserved for satellite high rate data transmissions.

2410.000–2413.000 MHz: This portion of the band is reserved for FM repeater outputs (25 kHz spacing).

2413.000–2418.000 MHz: This portion of the band is reserved for high rate data transmissions.

2418.000–2430.000 MHz: The portion of the band is reserved for satellite operations with 2430.000–2438.000 MHz reserved for satellite high rate data transmissions.

2438.000–2450.000 MHz: This portion of the band is reserved for wide-band FM, fast scan TV, FMTV, and slow scan experimental operations.

3300–3500 MHz

Operation on this band is open to all those holding a Technician class license or above. Weak signal operations are centered around ±100 kHz of 3456 MHz.

3300.000–3456.000 MHz: This portion of the band is reserved for local options.

3456.000–3456.050 MHz: This portion of the band is reserved for EME operations.

3456.050–3456.100 MHz: This portion of the band is reserved for CW operations.

3456.100–3456.300 MHz: This portion of the band is reserved for CW and SSB operations.

3456.300–3456.400 MHz: This portion of the band is reserved for beacons.

5650–5925 MHz

Operation on this band is open to all those holding a Technician class license or above. Weak signal operations are centered around ±100 kHz of 5760 MHz.

5650.000–5760.000 MHz: This portion of the band is reserved for local options.

5760.000–5760.050 MHz: This portion of the band is reserved for EME operations.

5760.050–5760.100 MHz: This portion of the band is reserved for CW operations.

5760.100–5760.300 MHz: This portion of the band is reserved for CW and SSB operations.

5760.300–5760.400 MHz: This portion of the band is reserved for beacons.

10.000–10.500 GHz

Operation on this band is open to all those holding a Technician class license or above. Weak signal operations are centered around 10.368 GHz.

10.3683–10.3684 GHz: This portion of the band is reserved for beacons.

24.000–24.500 GHz

Operation on this band is open to all those holding a Technician class license or above. Weak signal operations are centered around 24.368 GHz.

47.000–47.200 GHz

Operation on this band is open to those holding a Technician class license or above. Few weak signal operators are experimenting on this band. Whatever operations exist are centered around 47.100 GHz.

75.500–81.000 GHz

Operation on this band is open to all those holding a Technician class license or above. However, no regular weak signal or any other amateur radio type operation is currently taking place on this band.

119.980–120.020 GHz

Operation on this band is open to all those holding a Technician class license or above. However, no regular weak signal or any other amateur radio type operation is currently taking place on this band.

142.000–149.000 GHz

Operation on this band is open to all those holding a Technician class license or above. However, no regular weak signal or any other amateur radio type operation is currently taking place on this band.

241.000–250.000 GHz

Operation on this band is open to all those holding a Technician class license or above. However, no regular weak signal or any other amateur radio type operation is currently taking place on this band.

300.000 GHz and above

Operation on these frequencies is open to all those holding a Technician class license or above. Except for the light spectrum of 474 THz (red laser) and 668 THz (blue laser), no regular weak signal or any other amateur radio type operation is currently taking place on these frequencies.

Calling Frequencies

Because activity is less frequent on the VHF+ frequencies than on HF frequencies, certain calling frequencies have been set up and agreed upon by a majority of operators on the VHF ham bands.

Whether you came from HF or repeater operation, or are a new ham, you may be familiar with FM calling frequencies. However, you may not be familiar with the weak signal calling frequencies. The purpose of the calling frequency is to establish a gathering point for operators to initiate contacts. Once a contact has begun, operators are urged to move off the frequency in order to keep it clear for others to start contacts. Unfortunately, sometimes during band openings one station will stay on the calling frequency and dominate all contacts—unfairly reducing other operators' chances.

A list of calling frequencies for weak signal and FM work for each band follows.

50 MHz

28.885 MHz: This frequency is used to report immediate propagation on 6 meters from locations around the world. QSOs are discouraged, especially during 6 meter band openings. Once a contact is established on this frequency, the operators are encouraged to QSY immediately to a vacant frequency, usually up or down 5, 10, or 15 kHz. 28.885 MHz is invaluable to the serious 6 meter DXer, because so much current information is exchanged continuously.

While there is no formal net control, peer pressure generally keeps the frequency clear of idle chatter.

50.110 MHz: This is the DX-to-DX calling frequency. Intra-country and intra-continental QSOs (that is, QSOs within one's own country or, in the case of North America and Europe, one's own continent) are highly discouraged on this frequency, as well as between 50.100 and 50.125 MHz.

50.125 MHz: This is the domestic calling frequency. Once contact is established on this frequency, operators are encouraged to QSY up the band.

50.400 MHz: This is the domestic AM calling frequency. Many old-time 6 meter operators began on AM and haven't forgotten their love for it. Operations on this frequency occur across the country. Because it's not often used, communications established on this frequency tend to stay on this frequency. Therefore, it's not necessary to QSY, unless, of course, the band is open and the frequency becomes busy.

52.525 MHz: This is the domestic simplex FM calling frequency. Once contact is established on this frequency, operators are encouraged to QSY to another simplex frequency, such as 52.490 MHz or 52.510 MHz.

144 MHz

3.818 MHz: This frequency is an unofficial calling frequency for meteor scatter scheduling. It is especially active during major shower peaks. Be courteous when using this frequency, as regular 75 meter operators also use it for nets and rag chewing. An alternative to this frequency is 3.843 MHz.

14.345 MHz: This is the international EME calling frequency. It is similar in nature to the 10 meter activity frequency for 6 meters. Except on weekends (see below), there is no formal net. Because 20 meters is an extremely busy and crowded band, always check to see if you have a clear frequency before beginning a QSO. Once you've made a contact, be courteous and keep your QSO to an absolute minimum.

144.200 MHz: This is the calling frequency for all weak signal contacts. Once contact is established on this frequency, operators are encouraged to QSY up or down the band.

144.340 MHz: This is the ATV coordinating frequency used by operators who generally operate on 439.25 MHz.

146.430 MHz: This is the ATV coordinating frequency used by operators who generally operate on 434.000 MHz.

146.520 MHz: This is the domestic FM simplex calling frequency. Once contact is established on this frequency, operators are encouraged to QSY to another simplex frequency. Simplex frequencies are defined as those starting on 146.415 MHz and are found every 15 kHz up the band to 146.595 MHz. Simplex frequencies also are found starting on 147.420 MHz and running every 15 kHz up the band to 147.585 MHz.

222 MHz

222.100 MHz: This is the calling frequency for all weak signal contacts. Because there's less activity on this band, it's not as necessary to QSY once contact has been established, except during a contest.

223.500 MHz: This is the domestic FM simplex calling frequency. Once contact is established on this frequency, operators are encouraged to QSY to another simplex frequency. Simplex frequencies are defined as those starting on 223.420 MHz and are found every 20 kHz up the band to 223.900 MHz.

432 MHz

14.345 MHz: This is the international EME calling frequency. It's similar in nature to the 10 meter activity frequency for 6 meters. Except on weekends (see below), there is no formal net. Because 20 meters is an extremely busy and crowded band, always check to see if you have a clear frequency before beginning a QSO. Once you have made a contact, be courteous and keep your QSOs to an absolute minimum.

432.100 MHz: This is the calling frequency for all weak signal contacts. Here again, because there is less activity on this band, it's not as necessary to QSY once contact has been established—except during a contest.

446.000 MHz: This is the domestic FM simplex calling frequency. Because there is less simplex activity on this band, it's not necessary to QSY once contact has been established.

902 MHz

903.100 MHz: This is the calling frequency for all weak signal contacts. Once again, because there's so little activity on this band, it's not necessary to QSY once contact has been established—except during a contest. *Note:* In some areas 902.100 MHz is used as the alternative calling frequency.

906.500 MHz: This is the domestic FM simplex calling frequency. Again, because there's little simplex activity on this band, it's not necessary to QSY once contact has been established.

1240 MHz

1294.500 MHz: This is the domestic FM simplex calling frequency. Because there's less simplex activity on this band, it's not necessary to QSY once contact has been established.

1296.100 MHz: This is the calling frequency for all weak signal contacts. Again, because there's less activity on this band it's not as necessary to QSY once contact has been established, except during a contest.

2300 MHz

2304.100 MHz: This is the calling frequency for all weak signal contacts. Again, because there's less activity on this band it's not as necessary to QSY once contact has been established, except during a contest.

2305.200 MHz: This is the FM simplex calling frequency. Again, because there's less activity on this band, it's not necessary to QSY once contact has been established—except during a contest.

3300 MHz

3456.100 MHz: This is the calling frequency for all weak signal contacts. Again, because there's less activity on this band it's not necessary to QSY once contact has been established, except during a contest.

5650 MHz

5760.100 MHz: This is the calling frequency for all weak signal contacts. Again, because there's less activity on this band it's not necessary to QSY once contact has been established, except during a contest.

10 GHz

10.3681 GHz: This is the calling frequency for all narrowband (CW, SSB) activity.

A final note on calling frequencies: Most contests disallow contacts made on FM calling frequencies below 148 MHz. Also, some contests disallow domestic contacts made within the 6 meter DX window. My friend, Gordon West, WB6NOA, once wrote me decrying the activities of one "big gun" contest station that "parked" on 144.200 MHz during most of the ARRL June VHF Contest. Gordon felt it was a shame that many new hams' first experience with excellent propagation was marred because one operator inconsiderately monopolized a "calling frequency." In the situation in Gordon's example, the "big gun" could easily have been heard at 144.190 MHz or some other frequency close to the calling frequency.

Gordon also shared a story with me that sheds some light on this problem, by making a comparison between the calling frequency and surfing.

Some time ago one of the TV news magazine programs featured a surfing spot at Newport Beach known as the Wedge. Because of a jetty that right-angles the beach, waves created during a southerly swell in the surf bounce off it and form a wedge-like wall of water that can, as Gordon put it, " . . . catapult a body (surfer or fool!) down the face of the wave for a death-defying ride that lasts, at most, around 8 seconds."

From the reports, it's quite a ride. The news magazine program reported that it's so popular that a number of guys plan their days around weather reports favoring the right conditions for the Wedge. In fact, one report has it that a couple of guys carry pagers, just so they can be paged to the surf. Well, I'm sure you're beginning to see the similarity.

To continue, the Wedge is so confined that only a few people can ride it comfortably at a time. Unfortunately, there's a minority of surfers who want to ride it every time, blocking others who also want to experience the thrill of the ride.

I think you can see the comparison. The calling frequencies are only wide enough for a few to operate on them at a time. If the same people are on a calling frequency all the time, others can't even make a contact and move off the calling frequency.

So, the next time the surf's up (or the band is open) remember to share. A calling frequency is just that—a place to meet and establish contact. As a courtesy to your fellow VHF+ operators, please QSY to a clear frequency nearby once you establish contact on a VHF+ calling frequency.

Nets

Operation on the VHF+ frequencies sometimes requires coordination on the HF frequencies. Part of this coordination includes net operations. What follows is a list of the regular nets that will help you make more contacts on the VHF+ frequencies. For more listings, consult the *ARRL Net Directory*, available from the League for $2, including shipping, as of this writing.

3.818 MHz: Central States VHF Society Net. This is not exactly a net, but it meets every Sunday night, at 8:30 p.m., Central Time. Its structure is loose, with no net control. It is primarily for members of the Central States VHF Society. However, anyone is free to visit with anyone else on the frequency, as long as it's not busy. Obviously, operators wishing to conduct business that they anticipate will take some time are encouraged to move to another clear frequency. This frequency is also used by west coast stations almost nightly for coordinating their VHF+ activities, and it is used during meteor showers to coordinate schedules for meteor contacts.

3.840 MHz: AMSAT-NA Net This net meets throughout Tuesday evening, with sessions at 2100 Eastern time, then at 2100 Central time, and finally at 2000 Pacific time.

3.843 MHz: Midwest VHF Net. This net meets every Monday night at 9:00 p.m., Eastern Time. The net control stations are Brian, WA8MZQ, and Hal Perry, KC4YO. Check-ins are taken from all over the country. The net usually lasts for two hours. During the winter months when conditions are favorable, check-ins from the west coast are taken around 10:30 p.m., Eastern Time.

14.282 MHz: AMSAT International Net. This net meets Sunday at 1800 UTC.

14.345 MHz: EME Nets. These nets meet every Saturday and Sunday at 1600 and 1700 UTC, first the 432 MHz net and then the 144 MHz net.

18.155 MHz: AMSAT International Net. This net meets Sunday at 2300 UTC.

21.280 MHz: AMSAT International Net. This net meets Sunday at 1900 UTC.

Clubs, Newsletters, and Other Sources of Information

Following are some of the clubs, newsletters, and other resources that will prove of value to the VHF+ enthusiast. Prices of membership dues and subscriptions are sometimes given, but for the most recent information, and for shipping and handling charges, check with the individual clubs and publishers.

6 Meters

SMIRK. In order to qualify for membership in the Six Meter International Radio Klub (SMIRK), you must work six members and collect their numbers. Send your log data to: Acting Secretary Pat Rose, W5OZI, P.O. Box 393, Junction, Texas 76849, along with dues of $6. The club's newsletter hadn't been published for a few years because of founding scretary Ray Clark, K5ZMS's job responsibilities.

"The 50 MHz DX Bulletin." This newsletter is an excellent source of worldwide 6 meter activity. It's published monthly by Victor Frank, K6FV, 12450 Skyline Blvd., Woodside, California 94062-4541. Subscription rate is $20 annually.

Southern California Six Meter Club. If you live in the southern California area and are active on 6 meters, then you should join the Southern California Six Meter Club. The organization has monthly meetings and maintains a repeater on 52.86 MHz (minus 500 kHz offset). They also hold nets on the repeater on Thursdays at 8:00 p.m. Pacific Time, on SSB on 50.150 MHz on Tuesdays at 8:00 p.m. Pacific Time, and on AM on 50.400 MHz on Sunday at 10:00 a.m. Pacific Time. They also hold transmitter hunts the first Saturday of every month on 50.300 MHz, FM simplex. Membership is $10 and includes their newsletter, "The Six Pack." If you're interested, contact the club at P.O. Box 10441, Fullerton, California 92635.

2 Meters

Sidewinders on Two (SWOT). This is a fairly active national organization that maintains a listing of 2 meter nets around the country. The organization publishes a monthly newsletter. To join, work two members and collect their numbers. Send your log data, along with dues of $10, to: Howard Hallman, WD5DJT, 3230 Springfield, Lancaster, Texas 75134.

"2 meter EME News." This newsletter is published by John M. Carter, KØIFL, P.O. Box 554, Union, Missouri 63084. A subscription is $12. John also maintains a listing of those who are active on 2 meter EME and a data base of who's on what grid locator throughout the VHF-UHF spectrum. Contact John for the prices of each listing.

135 cm

"220 Notes, the National 220 MHz Newsletter." This newsletter, which was published for so many years by Art Reis, K9XA, is in custodianship and being held by Burt Hicks, WA6MQV. Burt is looking for someone to take over as editor. If you're interested, contact him at 28221 Stanley Court, Canyon Country, California 91350. You may also call him at (805) 251-5558, or FAX him at (805) 251-5572.

70 cm

"432 and Above EME News." This newsletter is published by Al Katz, K2UYH. To subscribe, send at least six months of SASEs to Allen Katz, K2UYH, Electronic Engineering Department, Trenton State College, Trenton, New Jersey 08650-4700. Your subscription is free, as long as you keep Al supplied with sufficient SASEs (number ten envelopes). He does accept contributions to pay for overseas subscribers' postage.

Microwave

North Texas Microwave Society. Although their name places them in Texas, this society's membership is international. They publish a monthly newsletter, **"Feedpoint,"** and membership dues are $12 per year. Send your money to: Wes Atchison, WA5TKU, Rt. 4, Box 565, Sanger, Texas 76266.

Microwave Update. This is an annual conference held in various locations around the country in the fall. For more information, contact Al Ward, WB5LUA, 2306 Forest Grove Estates Road, Allen, Texas 75002, or call (214) 542-6817.

The ARRL UHF/Microwave Experimenter's Manual. A useful book for understanding microwave technology, this manual includes information on design and fabrication techniques, propagation, antennas, and more. It is available from the ARRL for $20.

Microwave Handbook, Vols. 1, 2, 3. Published by the Radio Society of Great Britain, these handbooks cover operating techniques, systems analysis, antennas, microwave semiconductors, tubes, construction tips, equipment design, microwave beacons and repeaters, test equipment, and more.. Each is available from the ARRL for $35.

General

Mount Airy VHF Club. This club serves the Maryland, Delaware, Pennsylvania, Washington DC, New Jersey, and New York areas. They call themselves "The Pack Rats" and publish a monthly newsletter entitled **"The Cheese Bits."** To subscribe, send $7 to: Bob Fischer, WB2YEH, 7258 Walnut Avenue, Pennsauken, New Jersey 08110.

The Eastern VHF/UHF Society. This club services the eastern seaboard and New England areas. Send an SASE to: Thomas J. Kirby, W1EJ, 1 Meadow Knoll, Pelham, New Hampshire 03076 for membership information.

The Central States VHF Society. This regional organization covers the Midwest and Southwest. It holds an annual convention somewhere in the region on or around the last weekend in July. Proceedings are published by the ARRL and are available from the League for $12 plus $3.50 shipping and handling. Inquiries regarding membership and mailing information should be directed to: Larry C. Hazelwood, W5NZS, P.O. Box 54437, Oklahoma City, Oklahoma 73154.

"The West Coast VHFer." This newsletter serves the West Coast and the adjacent eastern states. It's edited by Bob Cerasuolo, WA6IJZ, P.O. Box 685, Holbrook, Arizona 86025. Subscription rate is $14.

West Coast VHF Conference. This is an annual get-together held in California around the second or third week of May. For more information, contact Gracie Hastings, KK6CG, at P.O. Box 10441, Fullerton, California 92635, or call (714) 990-9203.

"Upper Midwest VHF/UHF Newsletter." This newsletter serves the upper midwest and central Canada. It's edited by Rich Westerberg, N0HJZ, 17500 Cherry Drive, Eden Prairie, New Mexico 55346. Subscription rates are $7 per year in the U.S. and $10 per year in Canada. Rich has a number of contributors and thus lots of reporting on VHF+ activities in that area of the country.

"Anomalous Propagation." This is the newsletter for the Midwest VHF-UHF Society, Inc. It's co-edited by Gerd Schrick, WB8IFM, and Robert French, N8EHA. Society membership is $6 per year. Make your check out to Gerd Schrick, and send it to 1729 East Central Avenue, Miamisburg, Ohio 45342.

"N.E.W.S. Letter." Appropriately named for its organization, the **North East Weak Signal Group**, the "N.E.W.S. Letter" began publication in the fall of 1993. Published by the newly formed N.E.W.S. Group, which is an outgrowth of the 1993 East Coast VHF conference, this newsletter reports on weak signal activity, primarily in the New England area. Membership dues are $10. For a membership application, send an SASE to Ron Klimas, WZ1V, 458 Allentown Road, Bristol, Connecticut 06010.

"The Rocky Mountain VHF+ Newsletter." This newsletter was started by Doug Allen, W2CRS, and Ron Galbraith, KD0DW. It covers VHF+ activity in the Rocky Mountain states. To subscribe, send 12 SASEs plus $5 to 14231 E. County Line Road, Longmont, Colorado 80501.

Compilation

Beyond Line of Sight. This book is a compilation of many *QST* articles. It's edited by Emil Pocock, W3EP. It is an invaluable resource for historical research into weak signal activity. It is available from the ARRL for $12.

Your VHF Companion. This basic guide to operating on VHF is available from the ARRL for $8.

Amateur Television

Amateur Television Quarterly. A publication devoted to ATV, and this is available for $18 per year. It's edited by Henry Ruh, KB9FO. For a subscription, contact Henry at 1545 Lee Street, Suite 73, Des

Plaines, Illinois 60018 (708-298-2269). Henry has also published a book called *ATV Secrets*, which is available for $11.50, plus shipping and handling.

Amateur Satellite Communications

The ARRL Operating Manual. Chapter 13 of this book, written by John Bloom, KE3Z, provides a comprehensive, easy-to-understand look at amateur satellite operations. *The ARRL Operating Manual* is available from the ARRL for $18.

Satellite Experimenter's Handbook. This excellent reference for the satellite operator was written by Martin Davidoff, K2UBC The *Handbook* covers several of the active satellites in detail. Included is information on telemetry formats, uplink/downlink frequencies, on-board power systems, and more. It is available from the ARRL for $20.

ARRL Satellite Anthology. This is a collection of the best satellite articles from recent issues of *QST*. It is available from the ARRL for $8.

Decoding Telemetry from Amateur Satellites. This book, by G. Gould Smith, WA4SXM, shows how to receive and decode telemetry signals. It is available from AMSAT, P.O. Box 27. Washington, DC 20044, (301) 589-6062, for $15.

The Amsat Journal. This publication is available from AMSAT, P.O. Box 27, Washington, D.C. 20044, (301) 589-6062. A subscription is $30 a year.

The OSCAR Satellite Report. This is available from R. Myers Communications, P.O. Box 17108, Fountain Hill, Arizona 85269-71108. The cost is $56 a year U.S. and $62 Canada.

Orbital Elements

W1AW transmits orbital elements for all active amateur radio satellites twice a week. Orbital elements are also available on AMSAT Information Nets as follows:

International: Sunday at 1800 UTC on 21.280 and 14.282 MHz.

International: Saturday at 2200 UTC on 21.280 MHz.

U.S. East Coast: Tuesday at 2100 EST/EDT on 3.840 MHz.

U.S. Central: Tuesday at 2100 CST/CDT on 3.840 MHz.

U.S. West Coast: Tuesday at 2100 PST/PDT on 3.840 MHz.

FM and Repeaters

The ARRL Handbook for Radio Amateurs. This resource contains information on analog and digital electronics theory, antennas, transmitters, receivers, processing equipment, and accessories. The technical aspects of FM repeaters are discussed in Chapter 14. It is available from the ARRL for $25.

The ARRL Operating Manual. This is the complete guide to amateur radio operating. Chapter 11, by Brian Battles, WS1O, discusses FM and repeater operations. It is available from the ARRL for $18.

The ARRL Repeater Directory. Published annually, this pocket-sized book lists 20,000 repeaters in the United States and Canada on 29, 50–54, 222–225, 144–148, 420–450, 902–928, and 1240 MHz and above, including FM voice, packet radio, and amateur television repeaters. It also contains operating tips, band plan charts, club listings, and more. It is available from the ARRL for $6.

EME

Introductory EME Paper. Paul Kelley, N1BUG, has prepared an excellent introductory paper on EME operation. You may obtain a copy by sending a large SASE, plus units of first-class postage, to Paul Kelley, N1BUG, Box 33, Elm Street, Milo, Maine 04463.

EME Newsletter. John Carter, KØIFL, puts out an excellent 2 meter EME newsletter. You can obtain a year's subscription by sending $12 to: John Carter, KØIFL, P.O. Box 554, Union, Missouri 63084.

432 MHz and Above Newsletter. Al Katz, K2UYH, has an outstanding 432 and above EME newsletter. Subscriptions can be obtained via SASEs. Send at least six month's worth of SASEs to Allen Katz, K2UYH, Electronic Engineering Department, Trenton State College, Trenton, New Jersey 08650-4700. A subscription to Al's newsletter is free as long as you keep him supplied with sufficient SASEs (number ten envelopes). He does accept contributions to pay for overseas subscribers' postage.

Packet

The Packet Radio Operator's Manual. Written by Buck Rogers, K4ABT, this book is a guide to packet operation, plus provides detailed hookups for dozens of radio/packet controller/computer combinations. It is available from CQ Communications for $15.95.

Getting Started in Packet Radio. This video helps demystify the world of packet radio and tells you how to get started. It is available from CQ Communications for $19.95.

Your Gateway to Packet Radio. Written by Stan Horzepa, WA1LOU, this book provides a thorough discussion of packet radio. It is available from the ARRL for $12.

AX.25 Amateur Packet-Radio Link-Layer Protocol. This book gives the inner workings of the AX.25 packet protocol. It is available from the ARRL for $8.

"Packet Status Register." This newsletter is published quarterly by Tucson Amateur Packet Radio (TAPR), P.O. Box 12925, Tucson, Arizona 85732, for $15 a year.

Your Packet Companion. This is a basic book for packet operation. It is available from the ARRL for $8.

SSB and CW

The VHF/UHF DX Book. A look at VHF/ UHF operating is edited by Ian White, G3SEK. It is available from the ARRL for $35.

Radio Auroras. Charlie Newton, G2FKZ, discusses communication via auroral propagation in the Radio Society of Great Britain book. It is available from the ARRL for $18.

VHF/UHF Manual. Information on the history of VHF/UHF communications, propagation, receivers, transmitters, antennas, and more. Published by the Radio Society of Great Britain. It is available from the ARRL for $30.

"The West Coast VHFer." This newsletter is published by Ham Tech, P.O. Box 685, Holbrook, Arizona 86025, (602) 524-3354. Twelve issues per year for $14.

Meteors

METEORS. Written by Neil Bone, this is a new book released by Sky Publishing Corporation as part of their *Sky and Telescope* "Observer's Guides" series. For the radio amateur, this book provides a great deal of insight into what a meteor is, and how the meteor and Earth collide to create the visual (and, in our case, the electronic) observations we experience. The book gives a brief history of meteor studies, and contains a season-by-season calendar of annual meteor showers and their characteristics. A few paragraphs are devoted to the amateur radio operator's interest in meteor scatter propagation. It is available

for $18.95 from Sky Publishing Corporation, P.O. Box 9111, Belmont, Massachusetts 02178-9918.

Handbook for Visual Meteor Observations. The International Meteor Organization's *Handbook for Visual Meteor Observations*, edited by Paul Roggemans and published by Sky Publishing Corporation, covers meteor showers extensively. Included are historical anecdotes of both major and minor showers. I use this book almost every month when preparing my column for *CQ* magazine. The handbook is available for $18.95 from Sky Publishing Corporation, P.O. Box 9111, Belmont, Massachusetts 02178-9918.

Maps

ARRL World Grid Locator Atlas. See Chapter 15 for further information. It is available from the ARRL for $5.

ARRL World Grid Locator Map. See Chapter 15 for further information. It is available from the ARRL for $1.

Transmitter Hunting

Transmitter Hunting. This book tells you how be involved in the hidden transmitter hunt from both sides, whether you are the hunted (fox) or the hunter. By Joseph Moell, KØOV, and Thomas Curlee, WB6UZZ, it is available from the ARRL for $19.

Computer Bulletin Boards

HamNET Forums. Check the HamNET forums on Compuserve (enter GO HAMNET).

WB4YSA BBS. This bulletin board contains many useful files for the VHF+ operator. It is accessible by dialing (704) 284-4854. Data rates up to 38.4 KB are acceptable, with N, 8, and 1 being the format.

Awards

CQ USA County Award Record Book. This book, originally produced for CQ's USA-CA County Award, can also be used for the VUCA award sponsored by SWOT (see Chapter 16 for more information). It is available for $2.00 from CQ Communications, 76 N. Broadway, Hicksville, New York 11801.

Beacons

Beacons are propagation indicators used primarily by the weak signal operators on the VHF+ frequencies as ways of detecting band openings. Generally, the beacon transmission consists of the call sign of the station and the location either given in the Maidenhead grid locator or as a name, such as a city, or both.

Most beacons transmit continuously. However, because of the close proximity of the beacon to the owner's station, sometimes the owner shuts down the beacon when he is on the air.

In the United States, FCC regulations require that unattended beacons be contained within certain portions of the band. However, attended beacons can be placed on whatever frequency the operator decides.

Beacon stations are usually low power—10 watts or less. Antennas run the gamut, being anything from a loop to a vertical to multi-element beams pointed in one particular direction.

Most beacon operators would like to know that their beacon is being received. Therefore, if you hear one, copy down the call sign and send a note or a QSL card to the owner of the beacon.

VHF+ operators also use commercial transmissions as beacons. Such lists of stations are available from various sources.

The beacons in the following list are active as of January 1994. The 6 meter beacons were taken from the records maintained by Harry Schools, KA3B, Tim Marek, NC7K, and Orin Beebe, VE7BEE, and are published here with their permission. The 2 meter and above beacons were taken from records maintained by Hal Perry, KC4YO, and are published here with his permission.

BEACONS AS OF JANUARY 1994

Frequency	Grid	Callsign	Comments
50.008	—		KØGUV —
50.023	—	HH2PR	—
50.025	—	6Y5RC	—
50.041	—	WA8KGG	—
50.048	DO65	VE6ARC	50W GP
50.048	—	WA6IJZ	10W Vert.
50.055	—	WA9FEF	—
50.060	EM79	WA8ONQ	2W Turnstile
50.060	—	WA8ONQ	1W
50.060	—	KH6EQI	—
50.062	FN31	K1NFE	15W Turnstile
50.062	FM19	W3VD	10W Vert. Dipole
50.063	EL87	N4PZ	400MW 1/4 Vert.
50.064	CN85	N7DB	30W Antenna N/A
50.065	EL59	W5VAS	1W Halo
50.065	DM79	WØIJR	20W 2 Halos
50.065	DM79	KAØCDN	20W 2 Halos
50.065	DM59	WØMTK/B	2W MBL "Vee" Ant.
50.066	DM24	WD7Z	—

Frequency	Grid	Callsign	Comments
50.0664	—	KD4LP	—
50.067	EM79	WB8IGY	2W Vert.
50.067	DM04	WA6IJZ	70W Vert.
50.067	—	W7US	—
50.068	CM87	K6FV	—
50.070	EM55	W4HHK	2W Dipole
50.070	FM06	WB4GJG	1W Ringo Vert. Ant.
50.070	EM64	KA4VEY	10W Vert.
50.070	FM29	KS2T	Power N/A Antenna N/A
50.070	CN85	WA7ECY	30W Vert.
50.070	DN91	WØBJ	5W Turnstile
50.070	EM13	WBØCGH	1.5W Halo
50.070	EM55	N4NW	40W Squalo
50.070	FN41	W2CAP	—
50.071	—	W9KFO	—
50.072	FN12	WA2IYM	15W Turnstile
50.072	FN13	KW2T	1/4W Squalo
50.072	BL01	KH6HI	80W Dipole
50.072	—	VE1CCP	—
50.073	FN65	VE9MS/B	—
50.074	EM15	WB5DSH	30W Halo Ant.
50.075	DM43	K7IHZ	20W Squalo Ant.
50.075	FM06	WB4WTC	10W 2 Loops
50.075	FK68	KP4EKG	10W Vert. Ant.
50.077	EM09	NØLL	20W 2 Halos
50.077	EL49	N5JM	3W Vert. Ant.
50.077	FN30	WB2CUS	1W 5-el Yagi Ant.
50.077	FN21	WB2CUS/3	3W Squalo Ant.
50.077	—	VE3DRL	—
50.0775	EN75	W8UR	2W Dipole
50.079	DM41	W6SKC/7	50W 5/8 Vert.
50.080	EL87	WB4OOJ	10W Vert.
50.080	FN31	W1AW	80W West Beam Ant. (see *QST*)
50.086	FN46	VE2STL	1.5W Dipole VE2YB
50.086	FN46	VE2TH	3W Dipole
50.088	—	VE1SIX	—
50.090	AK56	KJ6BZ	10W 6-el Yagi
50.090	—	WA6JRA	—
50.092	EM40	W5GTP	30W 3-el Yagi
50.095	—	K7IHZ	—
51.199	DN09	VF7BEE	50W Vert.
52.000	All	FM	DX Call Channel
59.750	—	TV Sound	Kilowatts
144.051	EM86QV	WD4GSM	200MW to Halo
144.275	EM66	W4RFR	Chirps
144.275	EL88	WB4OOJ	25W to Big Wheel Omni
144.275	FN20	KS2D	100MW to Dipole
144.276	FN12	W2RTB	10W to Egg Beater Omni
144.277	EN15	KAØEWQ	10W to 4-el Yagi S.E.
144.277	FN08	VE2TWO	20W to Big Wheel Omni
144.277	GN03	VE1SMU/SI	10W to Yagi West
144.280	EM55	N4MW	5W to Squalo
144.280	EN58	VE3TBX	25W to 2-el Yagi S.E.
144.280	FN76	VE1VHF	7W
144.281	FN13	KW2T	—
144.282	EL87	WD4LWG	10 sec. Tone + ID/min.
144.282	EM13	WBØCGH	3W to Halo Omni
144.283	FM15	WD4KPD	5W to Omni
144.287	EN64	KB8JI	10W to 3-el Yagi S.S.E.
144.290	FN23XC	WA1UMX	20W to Phased Big Wheel Omni
144.290	EM27	WØVD	FSK
144.290	GN03	VE1SMU/H	10W NBFM 4-el Yagi S.W.
144.291	FM19	W3VD	15W to Halo Omni
144.292	FN24	W9IP/2	40W to 5-el No. for Aurora

Frequency	Grid	Callsign	Comments
144.294	EN01	WBØQIY	80W ? to 3-el S.E.
144.295	EM79QR	WA8R/9	15W to Turnstile Omni
144.297	EM55	N4ZRW	4W to Monopole N/S
144.297	FN65	VE1MS/B	Not Continuous 5W to Omni
144.297	EL98	WB2CUS/4	—
144.299	EM79	WZ8D	10W to Yagi Not Continuous
145.298	GN03	VE1SMU	Sable Isl.
147.930	GN03	VE1SMU	Sable Isl.
222.016	EM12	K5BYS	25MW to Omni
222.052	FN31	WB2IEY	30W to Turnstile Omni
222.055	FN12	W2RTB	10W to Eggbeater
222.055	EM55	N4MW	.5W to Squalo Omni
222.055	GN03	VE1SMU/SI	10W to 5-el Yagi Pointing West
222.060	GN03	VE1SMU/H	10W 2 Bay Vert. Omni
432.001	EM56	K4EJO	.5W Horiz. El. 3500 ft.
432.074	EN10	WBØQIY	—
432.075	EM86QV	WD4GSM	600MW to Halo
432.300	EN63VB	W2UHI	10W Omni—FSK
432.300	GN03	VE1SMU/SI	10W to 8 ft. Dish Pointing West
432.303	EM66	W4RFR	—
432.303	FN13	KW2T	—
432.315	FM19	W3VD	7W to Horiz. Omni
432.315	FN41	W1UHE	QRT if W1UHE QRV
432.317	FM07	WA4PGI	—
432.325	EM12	K5BYS	100MW to Halo
432.340	FN20	KB3NO	—
432.380	GN03	VE1SMU/H	15W 2-el Vert. S.W.
432.395	EM55	N4MW	10W 3/4 Horiz. Dipole Oriented E/W
902.285	EM55	N4MW	5W to Alford Slot Omni
902.325	EM12	K5BYS	4W to Turnstile
902.990	EM79	KA8EDE	5W to Yagi—Call for Directions
903.073	FN20	KB3NQ	—
903.075	EM86	K4EJQ	—
903.080	FN20	N3CXO	QRT if N3CX QRV—5W to Large Wheel
903.090	EN52	K3SIW	30W to Corner Refl. Aimed East
926.520	GN03	VE1SMU/H	NBFM to 1/4 Wave Vert.
1295.998	GN03	VE1SMU/H	20W to Hy-Gain Looper Aimed N.E.
1296.005	EM86	K4EJQ	—
1296.000	GN03	VE1SMU	Sable Isl.
1296.050	FM18	K3IVO	100MW NFSK, Drifts w/Temp. Changes
1296.070	EN52	K3SIW	20W to 15-el Log Periodic Yagi Aimed East
1296.073	EN52	K3SIW/9	—
1296.075	FN84	VE3SMU	—
1296.080	FN20	N3CX	QRT if N3CX QRV—1W to Slot
1296.180	EM26	WD5AGO	3W to Yagi to East
1296.202	FM07	WA4PGI	—
1296.260	FN12	KD5RO/2	1W to Halo
1296.325	EM12	K5BYS	300MW to Omni
1296.400	GN03	VE1SMU/H	5W to 2-el Horiz. West
2304.020	EN52	K3SIW	10W to Feedhorn Aimed West
2304.035	FN20	N3CX	QRT if N3CX QRV—4W to Slot
2304.050	EM86	K4EJO	—
2304.215	EM55	N4MW	4W to Slot
2304.325	EM12	K5BYS	40MW to Slot
2312.000	GN03	VE1SMU/H	1W to 4 ft. Dish Vert. West
3456.020	EN52	K3SIW	0.5W to Feedhorn Aimed West
3456.040	EM55	N4MW	20MW to Slot
3456.080	EM86	K4EJQ	—
3456.140	FN20	N3CX	2W to Alford Slot
3456.320	EM13SE	AA5C	+18dBm to Alford Slot
5759.950	EN52	K3SIW	5W to Feedhorn Aimed West

Index